Biochemistry Research Trends

science publishers
New York

Biochemistry Research Trends

Terpenes and Terpenoids
Charles A. Davies (Editor)
2022. ISBN: 978-1-68507-559-0

Circadian Rhythms and Their Importance
Rajeshwar P. Sinha, PhD, FSAB (Editor)
2022. ISBN: 978-1-68507-547-7 (Hardcover)
2022. ISBN: 978-1-68507-585-9 (eBook)

A Biochemical View of Antioxidants
Dorota Bartusik-Aebisher, PhD and David Aebisher, PhD (Editors)
2021. ISBN: 978-1-68507-151-6 (Hardcover)
2021. ISBN: 978-1-68507-295-7 (eBook)

Volatile Oils: Production, Composition and Uses
Sunita Singh, PhD (Editor)
2021. ISBN: 978-1-68507-186-8 (Hardcover)
2021. ISBN: 978-1-68507-241-4 (eBook)

An Introduction to Drug Carriers
Mohammad Ashrafuzzaman, D.Sc. (Editor)
2021. ISBN: 978-1-68507-148-6 (Hardcover)
2021. ISBN: 978-1-68507-157-8 (eBook)

Glycome: The Hidden Code in Biology
Dipak K. Banerjee, Ph.D. (Editor)
2021. ISBN: 978-1-53619-377-0 (Hardcover)
2021. ISBN: 978-1-53619-437-1 (eBook)

More information about this series can be found at
https://novapublishers.com/product-category/series/biochemistry-research-trends/

Charles A. Davies
Editor

Terpenes and Terpenoids

Sources, Applications and Biological Significance

nova
science publishers
New York

Library of Congress Cataloging-in-Publication Data

ISBN: 978-1-68507-559-0

Published by Nova Science Publishers, Inc. † New York

Contents

Preface

Terpenoids are the secondary metabolites of plant derivatives, mainly found in essential oils derived from higher plant sources. Terpenes are simple hydrocarbons, whereas terpenoids are modified terpenes containing different functional groups and oxidized methyl groups. This volume includes seven chapters examining topics related to terpenes and terpenoids. Chapter one describes mono- and sesquiterpenoids isolated from plants of the *Inula* L. genus and discusses biological activity of extracts and individual sesquiterpene lactones. Chapter two covers the nature of terpenoids together with their biological activities and their isolation from endemic species from the Asteraceae family growing in Algerian Sahara and used as medicinal plants. Chapter three introduces a catalyst system using Keggin heteropolyacid salts and hydrogen peroxide to epoxidize terpenic alcohols. Chapter four examines the phyto-active components of terpenoids as sources of insecticide. Chapter five describes how organic compounds containing functional groups like ketone, alcohol, and epoxide undergo rearrangement with sulfuric acid, yielding unexpected results. Chapter six focuses on alkoxylation of terpenes over heterogenous catalysts. Lastly, chapter seven presents a discussion about the importance of Brazilian flora for herbal medicine.

Chapter 1 - The review describes mono- and sesquiterpenoids isolated from plants of the *Inula* L. genus and discusses biological activity of extracts and individual sesquiterpene lactones.

Chapter 2 - The Algerian Sahara is an interesting area of biodiversity, which is a source of rich and abundant medical matter and an important reservoir for the research of new bioactive compounds with pharmacological potential. In Sahara many species from Asteraceae family are used in folk medicine for the treatment of diseases. Several studies demonstrated the antimicrobial, anti-inflammatory, antioxidant, anti-hyperlipidemia, hepatoprotection, cytoprotective, giving protection against cardiovascular disease and certain forms of cancer of Asteraceae species which contained a diversity of chemical compounds especially terpenoids, sesquiterpene lactones, triterpenoid saponin, steroids and polyphenolic compounds. Among these compounds, terpenoids are chemio-characteristic of Asteraceae family and having a potential application in pharmaceutical and cosmetic industries.

In this chapter the authors cover the nature of terpenoids together with their biological activities and their isolation from endemic species from Asteraceae family growing in Algerian Sahara and used as medicinal plants, namely: *Asteriscus graveolens*, *Launaea arborescens* and *Warionia saharae*

Chapter 3 - Terpenic alcohols are oxygenated derivatives from monoterpenes and can be easily obtained from natural raw materials such as extracts of wood and plants, and essential oils. Several oxidation products obtained from terpenic alcohols have pleasant organoleptic properties and are used as an ingredient in the synthesis of highly valuable compounds such as flavours, drugs, agrochemicals, and other fine chemicals. Controlling the reaction selectivity is one of the main challenges of these processes, mainly because these substrates can give carbonylic derivatives, isomers, and other products. Another problem in epoxidation reactions involving terpenic alcohols or other olefins is the use of stoichiometric quantities of toxic oxidants, which generate residues that should be disposed of into the environment. To circumvent this drawback, a green oxidant such as hydrogen peroxide can be used, however, it is always required the presence of a catalyst to do feasible these reactions. Noble metal catalysts are efficient in this process, nonetheless, their high price is a great disadvantage. An attractive alternative is the use of Keggin heteropolyacid (i.e., HPAs) or their derivatives salts as catalysts. In this work, the authors would like to highlight recent advances achieved in the development of processes for epoxides production from terpenic alcohols in Keggin heteropolyacid salts-catalyzed reactions using hydrogen peroxide as oxidant. Keggin HPAs are metal-oxygen clusters highly versatile as homogeneous or heterogeneous catalysts. On the other hand, hydrogen peroxide is a green oxidant that produces only water as a by-product. Therefore, the authors have developed a catalyst system using Keggin HPA salts and hydrogen peroxide to epoxidize terpenic alcohols (i.e., geraniol, nerol, linalool, α-terpineol, and β-citronellol). The authors will pay special attention to assessing the effects of main reaction parameters such as concentration, temperature, and stoichiometry of reactants. A comparison with traditional acid catalysts used in these reactions was also performed. Advances in the development of a new, efficient, and environmentally benign catalytic process to convert terpenic alcohols to epoxide were described. The authors hope that this work can significantly contribute to improving this important research field.

Chapter 4 - Natural products derived from plant sources have a wide range of biological applications. They are eco-friendly and mitigate or prevent many health ailments and environmental changes due to synthetic products. The long and persistent use of synthetic pesticides imposes a serious threat to non-target organisms and contaminates soil and groundwater. The plant-based products are bio-degradable in nature and could be considered as an alternate means of pesticides. Terpenoids are the secondary metabolites of plant derivatives, mainly found in essential oils derived from higher plant sources. They act on insects in various ways, mainly as chemical defenses, insect repellents, attractants, insecticides, and anti-feedents. Some are reported to influence the growth and development of some insects, which further leads to mortality and reproduction. Being volatile and lipophilic in nature, terpenoids could be used as fumigants which easily penetrate the insect body and might interfere with physiological, biochemical, and behavioural functions. Some terpenoids are phytotoxic to plants. This toxicity could be reduced by chemical modification of the parent compound. However, some of the derived compounds after chemical modification show enhanced pesiticidal activity. The understanding of the chemical nature of the compound and the influence of their physicochemical parameters are essential to predict the toxicological aspects of terpenoids and their insecticidal property.

Chapter 5 - The organic compounds containing functional groups like ketone (saturated, unsaturated), alcohol (primary, secondary, tertiary), epoxide etc. undergo rearrangement with sulfuric acid, yielding unexpected products. Alkyl groups attached to the aromatic ring migrate with sulfuric acid from one position to another. Several N-phenyl-hydroxyamines rearrange with sulfuric acid to amino phenols. The oximes with sulfuric acid furnish substituted amides and lactams.

Chapter 6 - Monoterpenes are renewable raw material. They can be used in the pharmaceutical, cosmetic and food industries as active components of drugs and ingredients of fragrances. Alkoxylation reactions of terpenes are important routes to obtain compounds with high commercial value. Traditionally, homogeneous acid catalysts (sulfuric acid) have been used in these reactions. However, these homogenous catalysts have some disadvantages such as, its separation from reaction mixture, they cannot be reused. The solid acids have many advantages over liquid acids. They do less harm to the environment and do not have corrosion or effluent disposal problems. They are reusable and easy to be separated from liquid products. This review focuses on alkoxylation of terpenes over heterogeneous catalysts.

Chapter 7 - This chapter presents a discussion about the importance of Brazilian flora for herbal medicine. Ethnomedical knowledge was possessed by different Brazilian native groups and consisted of complex practices, especially with the use of native herbs. Terpene and terpenoid compounds are natural products isolated from different natural sources. It consists of five-carbon isoprene units, which are assembled from many isoprene units. Terpenes are simple hydrocarbons, while terpenoids are modified terpenes with different functional groups (ketones, alcohols, esters, carboxylic acids, aldehydes) and oxidized methyl groups moved or removed at various structural positions. Terpenes and terpenoids are key components of essential oils. Brazil is considered the country with the greatest biodiversity in the world. According to the Brazilian Ministry of Environment, at least 103,870 animal species and 43,020 plant species are known. These species are distributed in six biomes with different characteristics and vegetation: Amazon Rainforest, Cerrado, Atlantic Forest, Caatinga, Pantanal, and Pampa. The wisdom of Brazilian natives transcends historical time and extermination to offer hope, especially in times of pandemics.

Chapter 1

Terpenoids of *Inula* L. Genus Plants

Rimma F. Mukhamatkhanova[*]**,**
Dilnoza E. Dusmatova and Ildar D. Sham'yanov
The Laboratory of the Chemistry of Terpenoids and Phenolic Compounds,
The Institute of the Chemistry of Plant Substances
AS RUz, Tashkent, Uzbekistan

Abstract

The review describes mono- and sesquiterpenoids isolated from plants of the *Inula* L. genus and discusses biological activity of extracts and individual sesquiterpene lactones.

Keywords: terpenoids, *Inula* L. genus, plant, biological activity

Introduction

Currently the interest in phytopreparations is growing due to their significant share in the world of pharmaceutical market. The main reason for the interest in herbal preparations is their wide range of pharmacological activity, the absence or minimal toxicity and the possibility of their long-term application in a number of chronic diseases. Medicinal plants are a renewable source of their production. A large share of medicinal plants of Eurasia grow in Central

[*] Corresponding Author's E-mail: rimmaraya@rambler.ru.

In: Terpenes and Terpenoids
Editor: Charles A. Davies
ISBN: 978-1-68507-559-0

Asia, among which plants of the family Asteraceae known as one of the largest families of the plant world.

Phytochemical studies of medicinal flora create the basis for the development of the pharmaceutical industry and make it possible not only to reduce the shortage of drugs, but also to contribute to the development of new, environmentally friendly and unique, export-oriented therapeutics. In this regard, the chemical study of plants producing biologically active compounds is one of the priority directions in the field of phytochemistry.

Terpenoids are one of the largest classes of secondary metabolites.

Natural terpenoids (or isoprenoids) are an extensive branched group of compounds of the secondary metabolism of plants, animals, algae and fungi. Isoprenoid biosynthesis occurs at all levels of living matter organization. The name of this class of compounds originates from the isoprene molecule, which consists of 5 carbon atoms and is an elementary link in the formation of these substances. The number of carbon atoms in the carbon skeletons of isoprenoids is a multiple of five. These compounds are known as isoprenoids or terpenoids.

Isoprene units can be attached to each other in various ways. The part of the molecule with geminal methyl groups is referred to as the "head," and the opposite part is referred to as the "tail". There are also isoprene units known to be attached as "head to tail," "head to head" and "tail to tail" (Semyonov, A. A., 2000; Plemenkov V. V., 2001).

The classification of terpenoids is based on the number of isoprene units included in the molecule (Semyonov, A. A., 2000; Plemenkov V. V., 2001) (Table 1):

Table 1. Classification of terpenoids

Terpenoid type	**Number of carbon atoms**	**The number of isoprene units**
Hemiterpenes	5	1
Monoterpenes	10	2
Sesquiterpenes	15	3
Diterpenes	20	4
Sesterterpenes	25	5
Triterpenes	30	6
Tetraterpenes	40	8
Polyterpenes	from 7000 to 300000	from 140 to 60000

Biogenesis of Terpenoids

Isopentyl pyrophosphate formed from acetate via mevalonic acid is considered as a precursor of the active five-carbon fragments in the biogenesis of terpenoids (Figure 1).

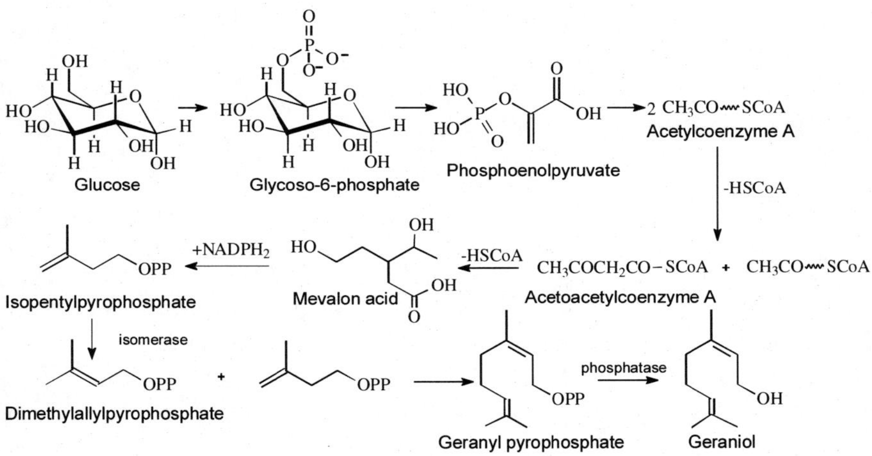

Figure 1. Biogenesis of terpenoids.

The resulting geraniol pyrophosphate is the ancestor of the most of existing monoterpenes. When isopentyl pyrophosphate is further added to it (an increase in the number of isoprene units), then farnesyl pyrophosphate to be formed, which in its turn is the ancestor of the sesquiterpenoid group, and so on. With the addition of another isopentyl pyrophosphate molecule, diterpenoids are formed; two molecules of farnesyl pyrophosphate and isopentyl pyrophosphate take part in the formation of tri- and tetraterpenes.

Terpenes are formed by combining isoprenoid units to form compounds such as geraniol (C10), farnesol (C15), geranylgeraniol (C20), and similar substances, and can be formed from these precursors by cyclization reactions and rearrangement mechanisms (Goodwin, T. W., 1964).

Plants of the genus *Inula* L. are rich sources of terpenoids.

Plants of the genus *Inula* L. - elecampane belong to the subtribe *Inulinae* Dumort. tribes *Inuleae*, subfamily Asteroideae, family Asteraceae. In the world flora, there are 595 species, including synonyms (http://www.the plantlist.org/), and in Central Asia there are more than 100 species of plants of this genus, of which 9 species grow in Uzbekistan

(Vvedenskiy, A. I., 1962). According to the conducted phytochemical studies, it was found that among the produced secondary metabolites of plants of this genus, mono- and sesquiterpenoids are dominant.

The production of preparations consisting of the active substances of terpenoids isolated from plants of the *Inula* L. genus, has been established in Russia, India, Poland, Belarus, Kazakhstan, Ukraine, USA, France, Switzerland.

Monoterpenoids of *Inula* L. Genus Plants

Classification

Most monoterpenoids are composed of two isoprene units attached head-to-tail. In nature, the carbon skeleton of monoterpenoids is formed by the reaction of isopentenyl pyrophosphate with dimethylallyl pyrophosphate (Figure 1) (Semyonov, A. A., 2000). As mentioned above, monoterpenoids can be subdivided into acyclic, monocyclic, bicyclic (Plemenkov V. V., 2001) and tricyclic. (Semyonov, A. A., 2000).

Table 2. Structural formulas of the main components of essential oil

Group name	Examples	Plant and literary source
	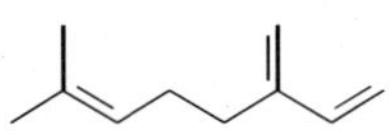 Myrcene	*I. oculus-christi* (Kiliç, O., 2014)
	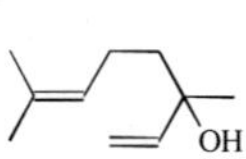 linalool	*I. oculus-christi* (Kiliç, O., 2014) *I. aschersoniana* (Todorova, M., Trendafilova, A., Dimitrov, D., 2017) *I. graveolens* (Al-Snafi, Ali Esmail, 2018) *I. viscosa* (Nasser, M., Housheh, S., Kourini, A., Maala, N., 2014) *I. helenium* (Stojanović-Radić, Z., et al., 2012) *I. cappa* (Priydarshi, R., Melkani, Anand B., Mohan, L., Pant Charu, C., 2015)
Acyclic	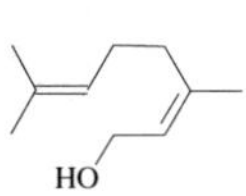 Nerol	*I. aschersoniana* (Todorova, M., Trendafilova, A., Dimitrov, D., 2017)

Group name	Examples	Plant and literary source
	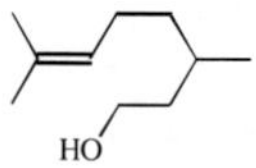 Citronellol	*I. helenium* (Karamenderes, C., Zeybek, U., 2000)
	 Linalyl acetate	*I. verbascifolia* (Ahmed, A., et al., 2003)
Acyclic	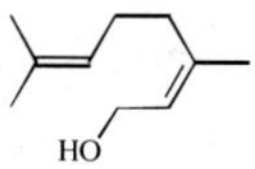 Geraniol	*I. aschersoniana* (Todorova, M., Trendafilova, A., Dimitrov, D., 2017) *I. viscosa* (Alalan, L., Al-Shammaa, I., Al-nouri, A. S., 2016)
Monocyclic	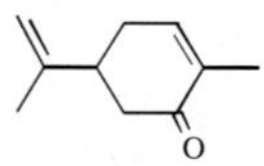 carvone	*I. graveolens* (Al-Snafi, Ali Esmail, 2018) *I. viscosa* (Nasser, M., Housheh, S., Kourini, A., Maala, N., 2014)
	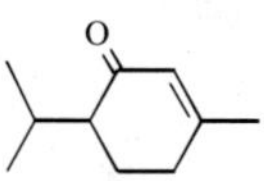 Piperitone	*I. graveolens* (Harzallah-Skhiri, F., Cheraif, I., Jannet, H. B., Hammami, M., 2005)
	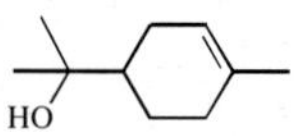 α-Terpineol	*I. aschersoniana* (Todorova, M., Trendafilova, A., Dimitrov, D., 2017) *I. britannica* (Todorova, M., et al., 2017) *I. helenium* (Karamenderes, C., Zeybek, U., 2000) *I. viscosa* (Laurentis, Onicolino de, Losacco, V., Milillo, M. A, Limpia, Lai, 2002)
Monocyclic	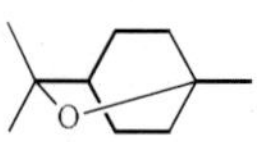 1,8-Cineole	*I. graveolens* (Al-Snafi, Ali Esmail, 2018) *I. oculus-christi* (Kiliç, O., 2014; Mitić, V. D., et al., 2019) *I. britannica* (Todorova, M., et al., 2017) *I. viscosa* (Laurentis, Onicolino de, Losacco, V., Milillo, M. A, Limpia, Lai, 2002) *I. macrocephala* (Kiliç, O., Baggi, E., 2013)
Monocyclic	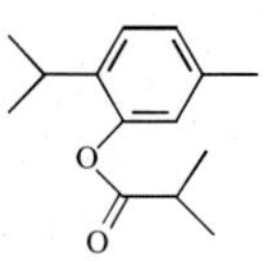 Thymil isobutyrate	*I. grandis* (Dusmatova, D. E., et al., 2019)

Table 2. (Continued)

Group name	Examples	Plant and literary source
	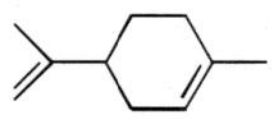 Limonene	*I. graveolens* (Al-Snafi, Ali Esmail, 2018) *I. oculus-christi* (Kiliç, O., 2014) *I. helenium* (Stojanović-Radić, Z., et al., 2012) *I. britannica* (Todorova, M., et al., 2017) *I. cappa* (Priydarshi, R., Melkani, Anand B., Mohan, L., Pant Charu, C., 2015)
	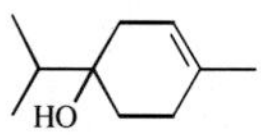 Terpinen-4-ol	*I. helenium* (Karamenderes, C., Zeybek, U., 2000) *I. britannica* (Todorova, M., et al., 2017) *I. cappa* (Priydarshi, R., Melkani, Anand B., Mohan, L., Pant Charu, C., 2015) *I. graveolens* (Boudouda, H. B., Kabouche, A., Aburjai, T., Kabouche, Z., 2013)
Bicyclic	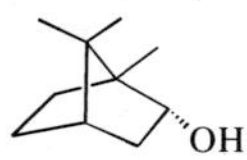 Borneol	I. oculus-christi (Kiliç, O., 2014) I. graveolens (Al-Snafi, Ali Esmail, 2018; Boudouda, H. B., Kabouche, A., Aburjai, T., Kabouche, Z., 2013; Harzallah-Skhiri, F., Cheraif, I., Jannet, H. B., Hammami, M., 2005; Karan, T., Yildiz, I., Aydin, A. Erenler, R., 2018; Afifi, F. U., Abu-Dahab, R., Kasabri, V. and Abaza, I. M., 2015; Beghidja, N., et al., 2014) I. macrocephala (Kiliç, O., Baggi, E., 2013)
	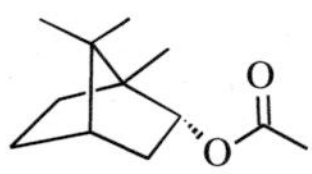 Bornyl acetate	I. oculus-christi (Kiliç, O., 2014) I. graveolens (Al-Snafi, Ali Esmail, 2018; Karan, T., Yildiz, I., Aydin, A. Erenler, R., 2018; Priydarshi, R., Melkani, Anand B., Mohan, L., Pant Charu, C., 2015; Beghidja, N., et al., 2014) I. crithmoides (Giamperi, L., et al., 2010)
	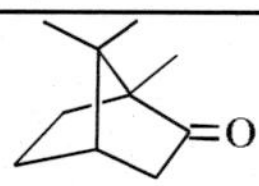 Camphor	I. oculus-christi (Kiliç, O., 2014) I. helenium (Stojanović-Radić, Z., et al., 2012; Hernandez-Ochoa, L., Vilarem, G., Mouloungui, Z., Medina-Gonzalez, Y., 2012) I. graveolens (Boudouda, H. B., Kabouche, A., Aburjai, T., Kabouche, Z., 2013) I. macrocephala (Kiliç, O., Baggi, E., 2013)
	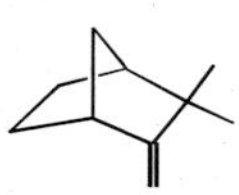 Camphene	*I. oculus-christi* (Kiliç, O., 2014) *I. graveolens* (Al-Snafi, Ali Esmail, 2018; Boudouda, H. B., Kabouche, A., Aburjai, T., Kabouche, Z., 2013; Harzallah-Skhiri, F., Cheraif, I., Jannet, H. B., Hammami, M., 2005; Karan, T., Yildiz, I., Aydin, A. Erenler, R., 2018; Afifi, F. U., Abu-Dahab, R., Kasabri, V. and Abaza, I. M., 2015) *I. macrocephala* (Kiliç, O., Baggi, E., 2013) *I. crithmoides* (Giamperi, L., et al., 2010) *I. helenium* (Stojanović-Radić, Z., et al., 2012; Hernandez-Ochoa, L., Vilarem, G., Mouloungui, Z., Medina-Gonzalez, Y., 2012)

Group name	Examples	Plant and literary source
Bicyclic	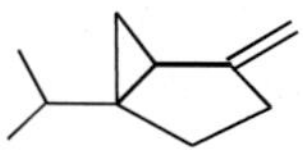 Sabinene	*I. oculus-christi* (Kiliç, O., 2014) *I. helenium* (Stojanović-Radić, Z., et al., 2012) *I. cappa* (Priydarshi, R., Melkani, Anand B., Mohan, L., Pant Charu, C., 2015) *I. crithmoides* (Giamperi, L., et al., 2010) *I. graveolens* (Afifi, F. U., Abu-Dahab, R., Kasabri, V. and Abaza, I. M., 2015)
	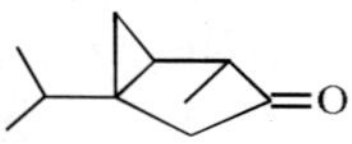 α-Thujone	*I. helenium* (Stojanović-Radić, Z., et al., 2012) *I. graveolens* (Boudouda, H. B., Kabouche, A., Aburjai, T., Kabouche, Z., 2013)
	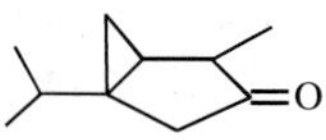 β-Thujone	*I. helenium* (Stojanović-Radić, Z., et al., 2012)

The main structural types of monoterpenoids are found in plants of the *Inula* L. genus are presented in Table 2.

Thus, plants of the genus *Inula* L. are rich sources of oxygenated monoterpenoids.

Composition of Essential Oils and Extracts

According to the available literature, the essential oils of plants *Inula britannica, I. aschersoniana, I. graveolens, I. viscosa, I. graveolens, I. oculus-christi, I. macrocephala, I. crithmoides, I. helenium, I. cappa, I. verbascifolia, I. cuspidata, I. grandis* were studied over the past 10 years.

The authors of (Todorova, M., et al., 2017) studied the essential oils of flowers and leaves of *Inula britannica* L., growing in Bulgaria, by capillary gas chromatography with a flame ionization detector (GC-FID) and gas chromatography-mass spectrometry. In total, 83 components were found in both oils, constituting 96.91% of the essential oil of flowers and 96.73% of the leaves, respectively. The terpenoids in these oils accounted for 57.85% and 77.28%, respectively, and contained sesquiterpenoids in majority. The main constituents of essential oils were viridiflorol (7.17% - 8.20%) and himachalol (3.45% - 8.71%), 6,10,14-trimethyl-2-pentadecanone (5.43% - 2, 95%), 13-tetradecanolide (3.93% - 4.87%) and 3-methyl-4-propyl-2,5-furandione (4.06% - 0.29%).

In the essential oil of *Inula aschersoniana* 45 components were found, accounting total of 92.9%. The oil contained fatty acids (55.2%) and alkanes (14.1%), followed by oxygenated monoterpenes (9.6%), sesquiterpenoids (7.1%) and aromatics (4.5%) (Todorova, M., Trendafilova, A., Dimitrov, D., 2017).

The next authors (Afifi, F. U., Abu-Dahab, R., Kasabri, V. and Abaza, I. M., 2015) studied the composition of *Inula graveolens* L. essential oils obtained by hydrodistillation and solid-phase microextraction by GC and GC-MS. The oxidized monoterpenes had the highest content in the hydrodistilled essential oil (87.32%), which consisted of bornyl acetate -the main component found in this fraction (70.58%). Both oxygenated monoterpene borneol (11.36%) and bornyl acetate (58.30%) remained as major components, while camphene was found as the major hydrocarbon monoterpene (16.58%) in the essential oil obtained by the solid phase microextraction. In the essential oil obtained by hydrodistillation method 16 components were identified, accounting for 98.21% of the total oil content. Oxygen-containing monoterpenoids accounted for 91.87% of the hydrodistilled oil, with the main component found in this fraction being bornyl acetate (70.58%). The monoterpene hydrocarbon fraction (2.57%) was represented mainly by camphene (1.97%). Sesquiterpenoids accounted for 2.74% of the total essential oil content in which the major was caryophyllene oxide (1.82%). The remaining identified sesquiterpenoids accounted for less than 0.5% of the total.

Analysis of *Inula graveolens* essential oil by GC and GC-MS methods, growing in Algeria, showed the presence of 47 components. The main components were: bornyl acetate (69.78%), borneol (4.25%), caryophyllene oxide (5.7%), mentadiene-8-ol (2.10%), chamigrene (2.9%) and β-selinene 0.95% (Beghidja, N., et al., 2014).

Essential oil of fresh aerial parts of *Inula graveolens* (L.) Desf. (Compositae), obtained by hydrodistillation in a Cleavenger-type apparatus was analyzed by GC-MS. In total 30 compounds were identified, constituting of 89.9% of the essential oil with isobornyl acetate (50.8%), borneol (18.3%) and τ-cadinol (6.2%) as the main components (Boudouda, H. B., Kabouche, A., Aburjai, T., Kabouche, Z., 2013).

Inula graveolens L. essential oil purchased from Huiles (Mandry, Corsica) was analyzed by GC, GC-MS and ^{13}C NMR methods. The main components were bornyl acetate (43.3%), borneol (26.2%) and camphene (7%) (Alalan, L., Al-Shammaa, I., Al-nouri, A. S., 2016).

The authors (Kiliç, O., 2014) studied the components of essential oils of the aerial parts of *Inula graveolens* Desf. and *Inula oculus-christi* L., grown in Turkey, by solid phase microextraction/GC-MS. 36 and 37 components were identified, representing 96.9% and 97.4% oil, respectively. The main compounds of *I. graveolens* are 1,8-cineole (22.4%), borneol (20.4%), and α-cadinol (11.8%), while for *I. oculus-christi* it is bornyl acetate (21.3%), p-cymene (16.6%) and β-pinene (14.8%).

The chemical composition of the leaves of *Inula viscosa* (L.) was analyzed by semi-preparative HPLC followed by GC-MS. The analysis led to the identification of 68 components, from which 6 terpenoids were isolated and identified: limonene, α-pinene, caryophyllene, α-terpenine, nerryl acetate, geraniol (Alalan, L., Al-Shammaa, I., Al-nouri, A. S., 2016).

The main components of the essential oil of the aerial part of *Inula macrocephala* Boiss. are borneol (26.4%), β-caryophyllene (15.3%), p-cymene (10.2%) and bornyl acetate (8.9%). The essential oil yield was about 0.13 ml/100 g of the plant with 35 components identified (Kiliç, O., Baggi, E., 2013).

The chemical composition of the essential oil obtained from the aerial parts of *Inula crithmoides* L. showed the presence of 22 identified components, the majority of which are *p*-cymene (30.1%), 1-methyl-ethyltrimethylbenzene (18.7%), scopoletin (15.3%) and α-pinene (13.1%). Essential oil has antioxidant activity (Todorova, M., et al., 2017).

Analysis of the essential oil of the wild plant *Inula cappa* collected in the Central Himalayan region (India) by capillary GC and GC-MS methods, showed that out of 90.2% of the total constituents of the oil, almost 50% were sesquiterpene hydrocarbons β-caryophyllene (27.5%), cis-dihydromayuron (6.7%), β-bisabolene (6.5%) and (E)-β-farnesene (5.6%) (Priydarshi, R., Melkani, Anand B., Mohan, L., Pant Charu, C., 2015).

The yield of *Inula helenium* essential oil obtained by hydrodistillation using the Clevenger apparatus was below 0.01% (v/w). Thus 91 compounds were identified accounting for 82.4%. The main components of the oil were caryophyllene oxide (14.7%), eudesma-5.11(13)-diene-8.12-olide, alantolactone (6.7%), isoalantolactone (3.8%) and aromadendrene oxide (3,3%). It was shown that the aerial part of *Inula helenium* also contains eudesmanolides, as well as the roots of this plant (Boudouda, H. B., Kabouche, A., Aburjai, T., Kabouche, Z., 2013).

The composition of the hexane fraction of the alcoholic extract of the roots of *Inula grandis* showed the presence of mainly sesquiterpenoids of the eudeman type and only one monoterpenoid thymyl isobutyrate (Dusmatova, D. E., et al., 2019).

To summarise, plants of the genus *Inula* L. are rich in oxygenated monoterpenoids such as borneol and bornyl acetate, as well as sesquiterpenoids such as caryophyllene and eudesmane.

Sesquiterpene Lactones of *Inula* L. Genus Plants

Sesquiterpenes are the largest group of all natural terpenes described (Plemenkov V. V., 2001; Rybalko K. S., 1978; Fischer, N. H., Olivier, E. J., Fischer, H. D., 1979; Kagarlizkiy A. D., Adekenov S. M., Kupriyanov A. N., 1987). This group of terpenoids in plants of the genus *Inula* L. is mainly represented by sesquiterpene lactones that have a wide spectrum of biological activity (*Plants resourses,* 1993; *Plant resources of Russia and neighboring countries,* 1996). In this regard, this group is of great interest to medicine, agriculture and veterinary medicine.

The larger size of the sesquiterpene molecule in comparison with monoterpenes provide great opportunities for a variety of ring closure methods and therefore, their chemistry is more diverse. There are more than 300 structural types (Semyonov, A. A., 2000).

Sesquiterpene lactones are cyclic esters of γ-hydroxy acids that have a five-membered lactone cycle (Scheme 1).

To date, about 8000 sesquiterpene lactones are known (Macias, F. A., et al., 2013).

The group of sesquiterpenes also includes highly active compounds of flora and fauna, such as phytoalexins protecting plants from diseases, ichthyotoxins, that are highly toxic to fish, as well as sesquiterpene lactones, that have a high spectrum of biological activity (Krizman, B. A., Stanzo, V. V., 1990).

The classification of sesquiterpenes is similar to monoterpenes - according to the degree of cyclization of the carbon skeleton.

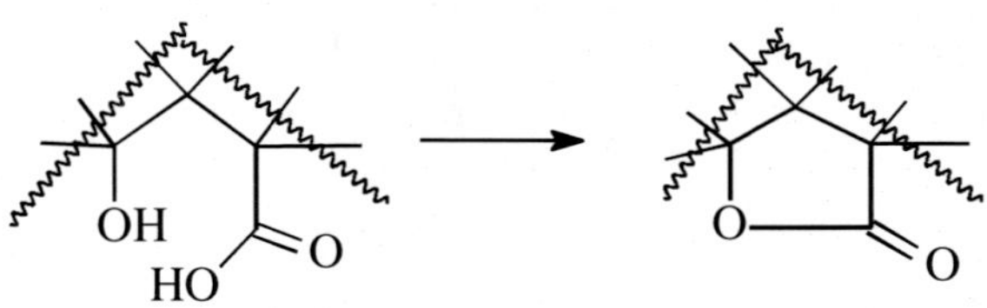

Scheme 1. Formation of γ-lactone.

Plants of the elecampane genus produce various structural types of sesquiterpene lactones: germacranolides, eudesmanolides, 1,10-seco-eudesmanolides, 4,5-secoeudesmanolides, guaianolides, pseudoguaianolides, xanthanolides, elemanolides, eremophylanolides, including dimeric lactones, and other types. Until 2015, lactones of plants of the genus *Inula* L. are described in the work (Seca, Ana M. L., Pinto, Diana C. G. A., Silva, Artur M. S., 2015).

Germacranolides

Germacranolides are bicyclic sesquiterpene lactones consisting of ten-membered and γ-lactone rings.

Table 3 shows examples of isolated germacranolides from plants of the *Inula* L. genus.

Germacranolides are the ancestors of the main types of sesquiterpene lactones (Mukhamatkhanova, R. F., 2018). Biosynthesis of germacranolides proceeds through germacrene A, formed from farnesyl diphosphate, followed by germacrane precursor in the presence of sesquiterpene cyclases being converted to guaianolides or eudesmanolides, depending on the position of the enzyme-substrate binding (Piet, D. P., Schrijvers, R., Frassen, M. C. R., De Groot, A., 1995).

Figure 2 shows a diagram of the formation of various types of lactones from germacrane precursor.

Due to the large number of compounds and structural variants of sesquiterpenoids assigning a sesquiterpenoid to a certain type depends not only on the carbon skeleton, but also the spatial structure of the molecule is taken into account.

The conformation of a molecule (from the Latin word conformatio – "shape, structure, arrangement") is the spatial arrangement of atoms in a molecule of a certain configuration, due to rotation around of one or more single sigma bonds.

The ten-membered carbocycle of sesquiterpene lactones of the germacrane type with two intracyclic double bonds at the C-1,10- and C-4,5-positions have unique features. Depending on the configuration of endocyclic double bonds, germacranolides are subdivided into 4 types, the structures and names of which are given below (Figure 3):

Table 3. Germacranolides of plants of the genus *Inula* L.

№	Sesquiterpene lactone	Structure	Source
1	Costunolide $C_{15}H_{20}O_2$ Mr 232 m.p. 106-107°C, $(\alpha)_D$ +128°	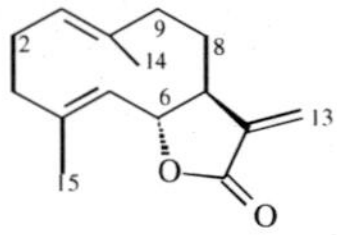	*I. racemosa* (Venkateswara, Rao G., et al., 2013)
2	Eupatolide $C_{15}H_{20}O_3$ Mr 248 m.p. 226-231°C $(\alpha)_D^{24,9}$ +49.75 (*c* 0.27, MeOH)	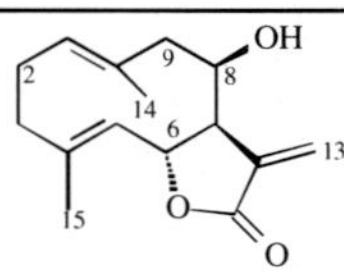	*I. japonica* (Qin, J. J., et al., 2010) *I. japonica* (Wu, X. D., et al., 2016)
3	Deacetylovatifolin $C_{15}H_{20}O_4$ Mr 228 m.p. 190-195 °C, $(\alpha)_D^{24,0}$ -26.71 (*c* 0.15, MeOH)	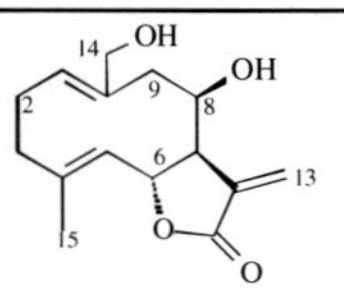	*I. lineariifolia* (Qin, J. J., et al., 2013) *I. japonica* (Wu, X. D., et al., 2016)
4	9β-Hydroxyparthenolide $C_{15}H_{20}O_5$ Mr 226	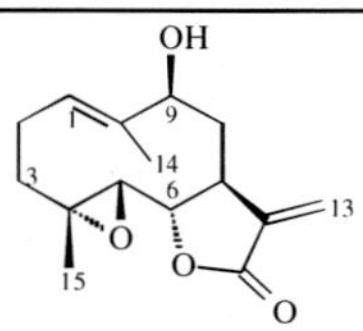	*I. candida* subsp. *candida, I. verbascifolia* (Michalakea, E., et al., 2019)
5	9β-{((2'*S*)-2'-Methyl-3'-oxobutanoyl)oxy}-parthenolide $C_{20}H_{26}O_6$ Mr 362	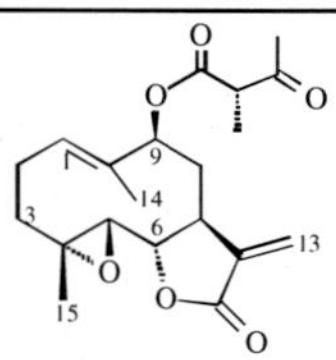	*I. candida* subsp. *candida* (Michalakea, E., et al., 2019)
6	9β-((3'-Hydroxyisovaleryloxy) parthenolide $C_{20}H_{28}O_6$ Mr 364	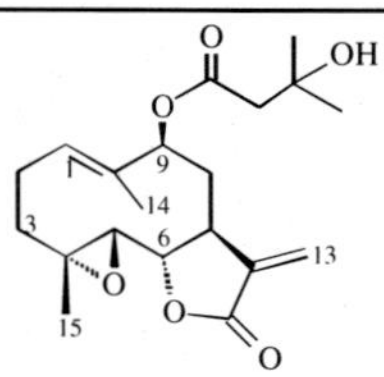	*I. candida* subsp. *candida* (Michalakea, E., et al., 2019)
7	Pterocaullin A $C_{20}H_{26}O_5$ Mr 346 m.p. 172–173°C	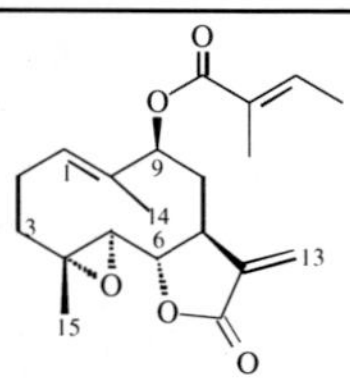	*I. pterocaula* (Zhu, S. L., et al., 2019)

№	Sesquiterpene lactone	Structure	Source
8	Pterocaullin B $C_{20}H_{26}O_5$ Mr 346 m.p. 159–161°C	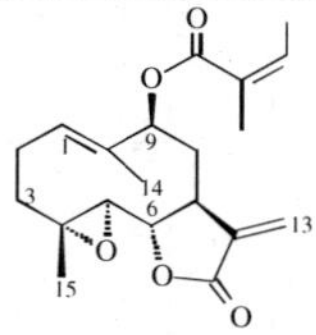	*I. pterocaula* (Zhu, S. L., et al., 2019)
9	Pterocaullin C $C_{21}H_{28}O_5$ Mr 360 m.p. 171–172°C	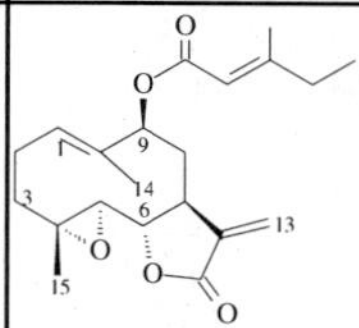	*I. pterocaula* (Zhu, S. L., et al., 2019)
10	Pterocaullin D $C_{20}H_{26}O_6$ Mr 362 m.p. 170–171°C	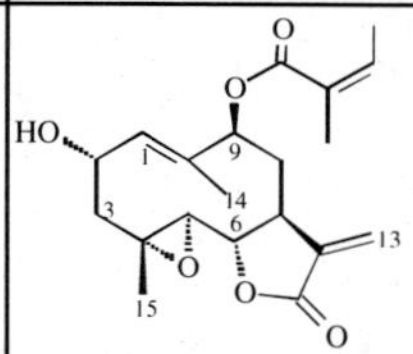	*I. pterocaula* (Zhu, S. L., et al., 2019)
11	Inuloxine A $C_{15}H_{20}O_3$ Mr 212 $(\alpha)_D^{25}$ 8,5 (*c* 1.9)	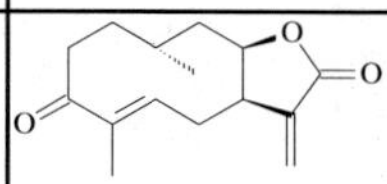	*I. viscosa* (Andolfi, A., et al., 2013)
12	Inuloxine B $C_{15}H_{22}O_3$ Mr 214 $(\alpha)_D^{25}$ + 49,9 (*c* 0.2)	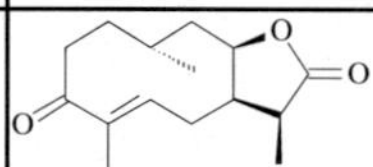	*I. viscosa* (Andolfi, A., et al., 2013)
13	4*E*-9β-Hydroxy-7α,10αH-germacra-4,11(13)-dien-12,6α-olide $C_{15}H_{22}O_3$ Mr 214 $(\alpha)_D^{20}$ + 69.6 (*c* 0.10, CH_2Cl_2)	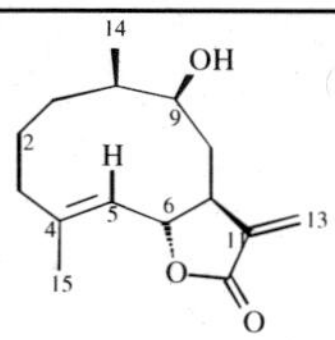	*I. wissmanniana* (Cheng, X. R., et al., 2013a)
14	(4*E*,6*R*,7*S*,9*S*,10*R*)-9-Angeloyloxygermacra-4,11(13)-dien-12,6-olide $C_{20}H_{28}O_4$ Mr 332 m.p. 115–125°C $(\alpha)_D^{20}$ + 90.0 (*c* 0.10, CH_2Cl_2) $(\alpha)_D^{20}$ +135.8 (*c* 0.10, MeOH)	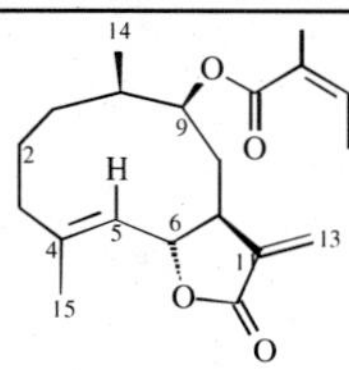	*I. wissmanniana* (Cheng, X. R., et al., 2013a; Cheng, X. R., et al., 2014)

Table 3. (Continued)

№	Sesquiterpene lactone	Structure	Source
15	4*E*-9β-Methacryloxy-7α,10αH-germacra-4,11(13)-dien-12,6α-olide $C_{19}H_{26}O_4$ Mr 318 $(\alpha)_D^{20}$ +77.1 (*c* 0.05, CH_2Cl_2)	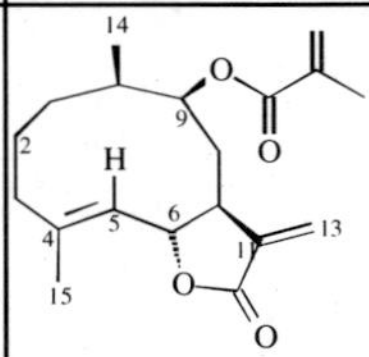	*I. wissmanniana* (Cheng, X. R., et al., 2013a)
16	4*E*-9β-Angeloyloxy-2α-hydroxy-7α,10αH-germacra-4,11(13)-dien-12,6α-olide $C_{20}H_{28}O_5$ Mr 348 m.p. 134–142°C, $(\alpha)_D^{20}$ +104.9 (*c* 0.10, CH_2Cl_2)	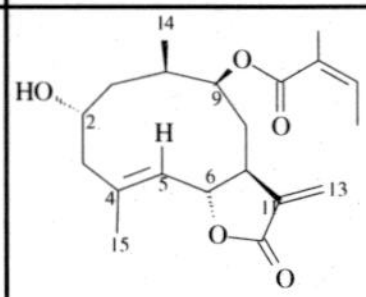	*I. wissmanniana* (Cheng, X. R., et al., 2013a)
17	(2*R*,4*E*,6*R*,7*S*,9*S*,10*R*)-2-Acetoxy-9-angeloyloxy-germacra-4,11(13)-dien-12,6-olide $C_{22}H_{30}O_6$ Mr 390 m.p. 157–160°C, $(\alpha)_D^{20}$ + 119.0 (*c* 0.10, CH_2Cl_2)	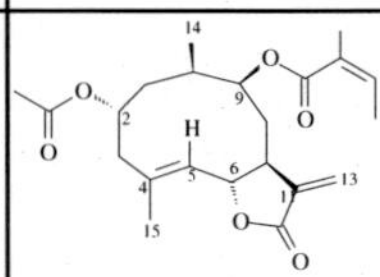	*I. wissmanniana* (Cheng, X. R., et al., 2013a)
18	(4*R*,5*R*,6*S*,7*S*,9*S*,10*R*)-9-Angeloyloxy-4,5-epoxy-germacra-11(13)-en-12,6-olide $C_{20}H_{28}O_5$ Mr 348 m.p. 167–175°C, $(\alpha)_D^{20}$ +61.6 (*c* 0.10, MeOH)	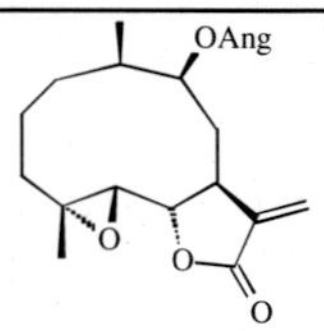	*I. wissmanniana* (Cheng, X. R., et al., 2014)
19	Ineupatolide D $C_{26}H_{36}O_9$ Mr 492	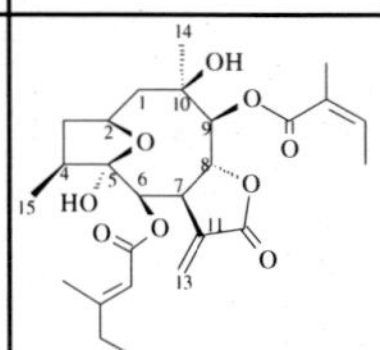	*I. cappa* (Wu, J. W., et al., 2017)
20	Ineupatolide E $C_{20}H_{26}O_7$ Mr 378 $(\alpha)_D^{20}$ +128.0 (*c* 0.1, $CHCl_3$)	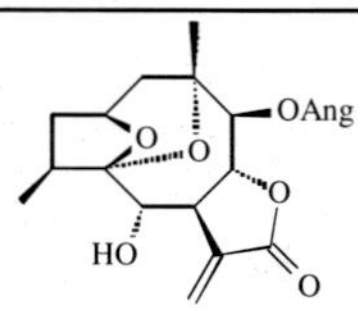	*I. cappa* (Wu, J. W., et al., 2017)

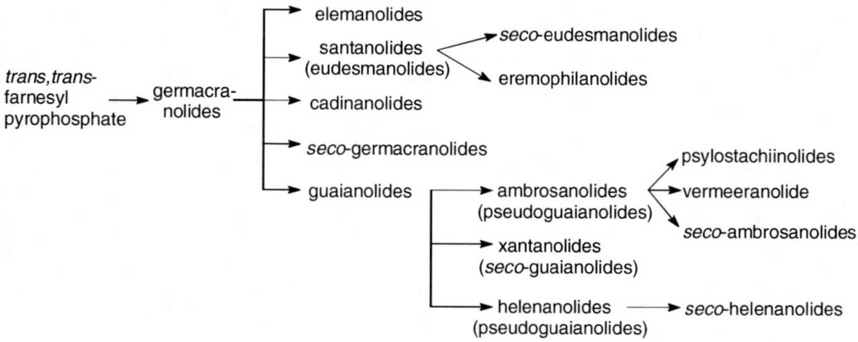

Figure 2. Formation scheme of various types of lactones from germacrane precursor.

E,E-germacrolides, for example, include lactones **1-10**, *E,Z*-heliangolides - lactones **11-12**, *Z,E*-melampolides - lactones **13-18**, *Z,Z*-germacranolides - lactones **19-20** (Table 3).

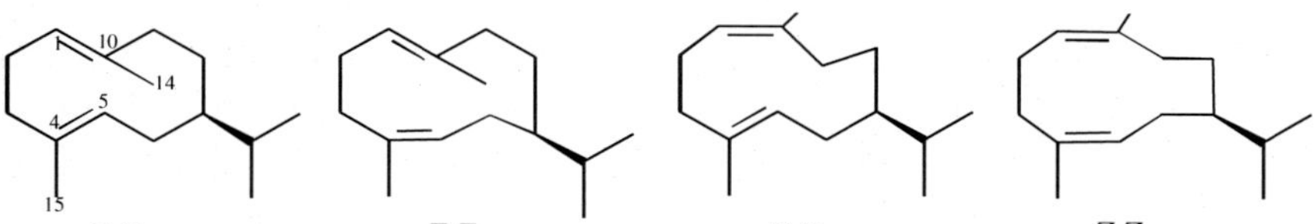

E,E- Germacrolide   *E,Z*- Heliangolide   *Z,E*- Melampolide   *Z,Z*- *Z,Z*-Germacranolide

Figure 3. Characteristic conformations of germacranolides.

Table 4. Types of germacranolide configurations

Types	**Double bonds**		**Types**	**Double bonds**	
	1,10	4,5		1,10	4,5
E,E	${}_{1}D^{14}$	${}^{15}D_{5}$	*E,Z*	${}_{1}D^{14}$	${}^{15}D^{5}$
	${}_{1}D^{14}$	${}_{15}D^{5}$		${}_{1}D^{14}$	${}_{15}D_{5}$
	${}^{1}D_{14}$	${}_{15}D^{5}$		${}^{1}D_{14}$	${}^{15}D^{5}$
	${}^{1}D_{14}$	${}^{15}D_{5}$		${}^{1}D_{14}$	${}_{15}D_{5}$
Z,E	${}^{1}D^{14}$	${}^{15}D_{5}$	*Z,Z*	${}_{1}D_{14}$	${}^{15}D^{5}$
	${}^{1}D^{14}$	${}_{15}D^{5}$		${}^{1}D^{14}$	${}_{15}D_{5}$
	${}_{1}D_{14}$	${}_{15}D^{5}$			
	${}_{1}D_{14}$	${}^{15}D_{5}$			

In the above-mentioned types *E,E*-germacrolides with *trans,trans*-configuration of double bonds in the ten-membered ring, *E,Z*-heliangolides with *trans,cis*-configuration and *Z,E*-melampolides with *cis,trans*-configuration of double bonds, 4 types of configurations are realized, and *Z,Z*-

germacranolides with *cis,cis*-configuration - 2 types of configurations (Table 4).

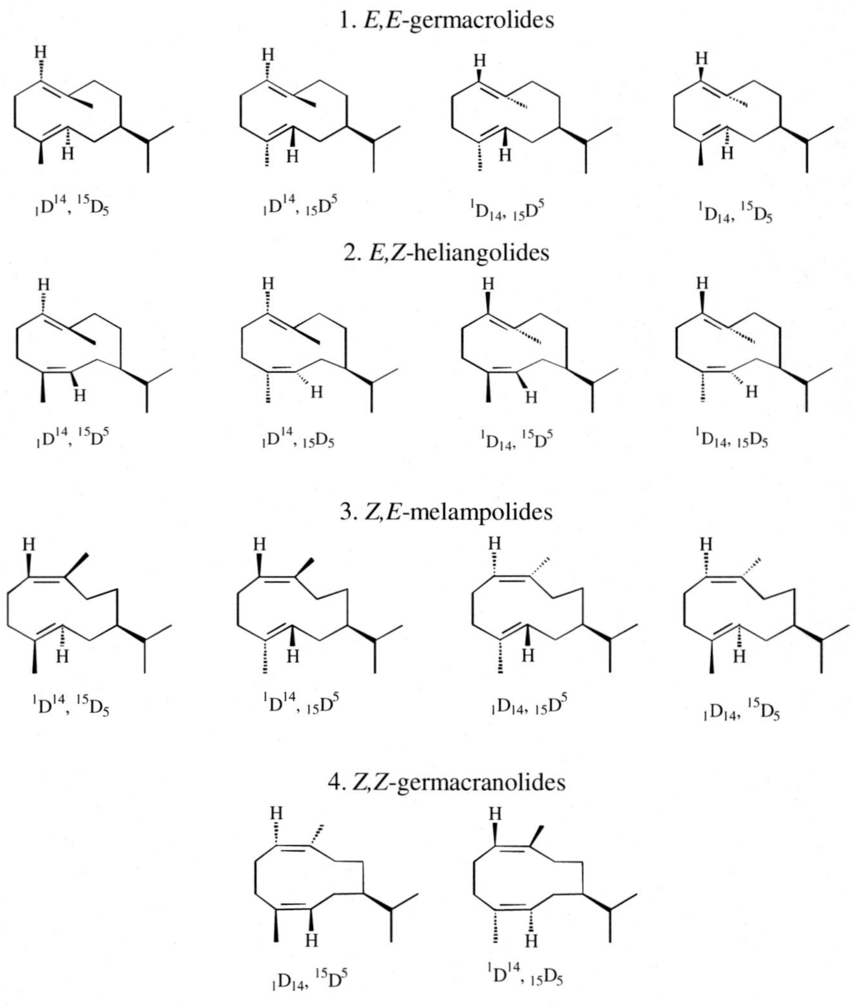

Figure 4. Relative intracyclic configuration of double bonds in germacranolides.

The prefixes *Z, E* and indexes in the D symbol indicate the orientation of the C14 and C15 atoms (methyl groups) and hydrogen atoms in the 1(10) and 4(5) double bonds relative to the rms plane of cyclodecadiene, with the superscripts of the D symbol pointing upwards and the subscripts of the D character go down.

The relative intracyclic configuration of double bonds of all four types of germacranolides is demonstrably shown below (Figure 4):

The above picture 4 shows that the prefixes *Z* and *E*, as well as the indices of the symbol D, clearly depict the configuration of methyl groups and hydrogen atoms at double bonds within the ten-membered carbocycle of germacrane lactones.

The conformation of germacranolides is influenced by the location and nature of the functional groups in the molecule, especially in the solid state of aggregation. An important role is played by intra and intermolecular hydrogen bonds. In solutions, molecules can accept any conformation, passing from one to another.

Table 5 lists the conformations of the cycles of natural plant germacranolides, established by X-ray diffraction analysis.

Based on the analysis of tabular data (Table 5), most germacranolides have a chair-boat decimal cycle conformation with the lactone cycle taking the form of an envelope or a half-chair.

Eudesmanolides

Eudesmanolides are sesquiterpene lactones consisting two six-membered and γ-lactone rings.

Eudesmanolides are classified as linear and non-linear, for example alantolactone and tauremisin, respectively (Figure 5).

The lactone cycle to the skeleton of the sesquiterpene lactone molecule can be *trans*- or *cis*-linked. The structure of the majority of lactones found in plants of the *Inula* L. genus is dominated by the *cis*-connection of the lactone cycle and the presence of an α,β-unsaturated γ-lactone group.

Figure 5. Linear and nonlinear eudesmanolides.

Table 5. Conformations of the cycles of natural plant germacranes established by X-ray diffraction

№	Compound	Conformation	Literature
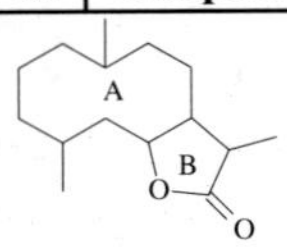			
1	Ugaferin	cycle A has chair-boat conformation	(Makhmudov, M. K., Tashkhodzhaev, B., Saidkhodzhaev, A. I., Ibragimov, B. T., 1990)
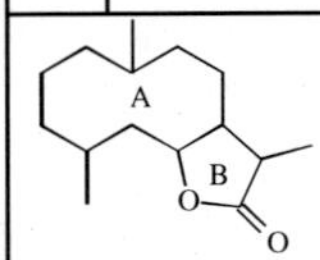			
2	Subchrysin	cycle A has chair-boat conformation, cycle B has half-chair conformation	(Turdybekov, K. M., et al., 1998a)
3	Mucrochin	cycle A has chair-boat conformation, cycle B has envelope conformation	(Izbosarov, M. B., et al., 1998)
4	Argolide	cycle A has boat-chair conformation, cycle B has half-chair conformation	(Turdybekov, K. M., et al., 1998b)
5	Desacetylajanolide A	cycle A has chair-boat conformation, cycle B has envelope conformation	(Kulyyasov, A. T., et al., 1999)
6	1β,10α-Epoxy-ajanolide A	cycle A has boat-chair conformation, cycle B has half-chair conformation	(Kulyyasov, A. T., et al., 1999)
7	Jurineolide	cycle A has chair-chair conformation, cycle B has twist conformation	(Adekenov, S. M., et al., 1991)
8	Mikagoyanolide (dimeric lactone)	cycle A has chair-chair, cycle A' has boat-chair, lactone cycles B and B' have between the envelope and the half-chair conformations	(Makhmudov, M. K., Tashkhodzhaev, B., Abduazimov, B. Kh., 1993)
9	Ketopelenolide A	cycle A has boat-boat conformation, cycle B has envelope conformation	(Makhmudov, M. K., Tashkhodzhaev, B., Zakirov, S. Kh., 1990b)

Table 6 shows examples of isolated eudesmanolides from plants of the *Inula* L. genus.

The conformation of eudesmanolides is determined by the conformations of the six-membered and five-membered rings. Five-membered cycles can take the envelope conformation or twist-conformation (half-chair) that quickly transforms into each other. In this case, all five corners of the envelope are alternately bent (pseudo-rotation) (http://www.chimfak.sfedu.ru/) (Figure 6).

Table 6. Eudesmanolides of the *Inula* L. genus plants

№	Sesquiterpene lactone	Structure	Sourse
1	Isoalantolactone $C_{15}H_{20}O_2$ Mr 232 m.p. 117-119°C (ethanol)	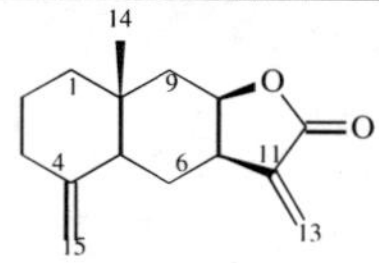	*I. helenium* (Ghedira, K., Goetz, P., Le, R. Jeune, 2011)
2	1-Epigranilin $C_{15}H_{20}O_4$ Mr 228 $(\alpha)_D^{20}$ +146.0 (*c* 0.1, MeOH)	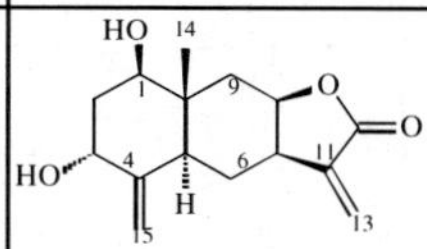	*I. britannica* (Fischedick, J. T., et al., 2013)
3	1β,6α-Dihydroxy-5αH-eudesma-4(15),11(13)-dien-12,8β-olide $C_{15}H_{20}O_4$ Mr 228 $(\alpha)_D^{20}$ +84.2 (*c* 0.1, MeOH)	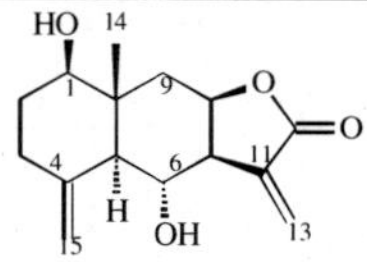	*I. britannica* (Zhang, X. F., et al., 2015)
4	5α-Hydroxyasperilin $C_{15}H_{20}O_4$ Mr 228	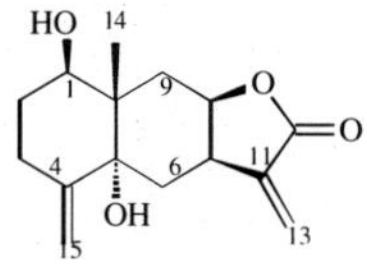	*I. japonica* (Qin, J. J., et al., 2010)
5	5α-Hydroperoxy-asperilin $C_{15}H_{20}O_5$ Mr 244	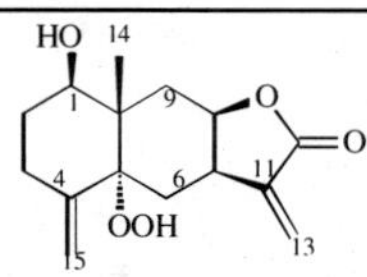	*I. japonica* (Qin, J. J., et al., 2010)
6	Alantolactone C15H20O2 Mr 232 m.p. 76-77°C (ethanol) (α)D +187.96°	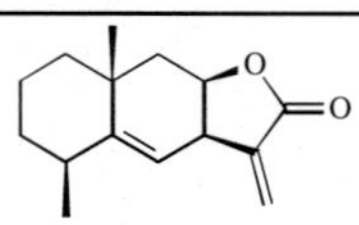	*I. helenium (Ghedira, K., Goetz, P., Le, R. Jeune, 2011)* *I. racemosa (Xu, L. W., Shi, Y. P., 2011)*
7	1β-Hydroxyalantolactone C15H20O3 Mr 212	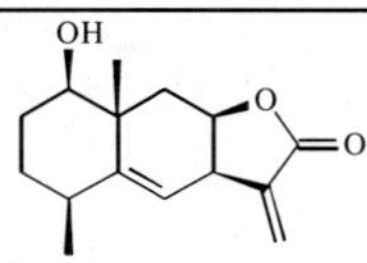	*I. japonica (Qin, J. J., et al., 2010)*

Table 6. (Continued)

№	Sesquiterpene lactone	Structure	Sourse
8	Inujaponin F $C_{17}H_{22}O_4$ Mr 290 $(\alpha)_D^{26,5}$ +218.62 (*c* 0.14, MeOH)	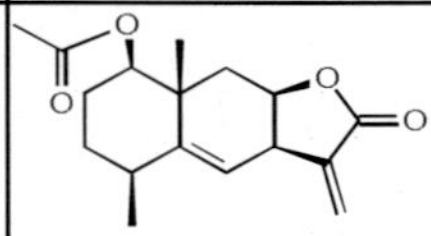	*I. japonica* (Li, X., et al., 2020)
9	JEUD-38 $C_{15}H_{18}O_3$ Mr 210	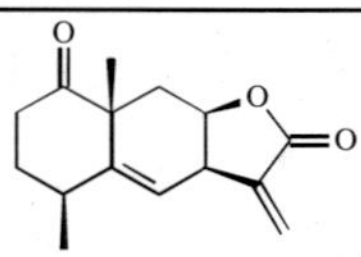	*I. japonica* (Trendafilova, A., Ivanova, V., Todorova, M., Aneva, I., 2017)
10	3β-Hydroxy-11α,13-dihydroalantolactone E $C_{15}H_{22}O_3$ Mr 214	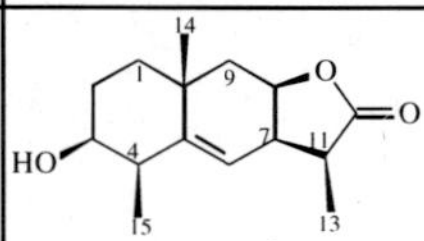	*I. leariifolia* (Qin, J. J., et al., 2013)
11	Racemosalactone C $C_{15}H_{22}O_3$ Mr 214	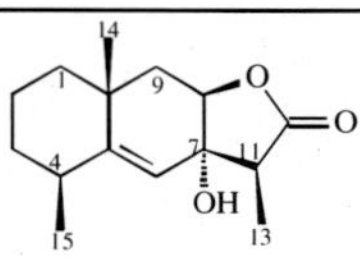	*I. rasemosa* (Ma, Y. Y., Zhao, D. G., Gao, K., 2013)
12	11α-Hydroxyeudesm-5-en-12,8β-olide $C_{15}H_{22}O_3$ Mr 214	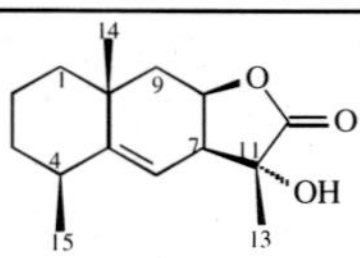	*I. racemosa* (Xu, L. W., Shi, Y. P., 2011)
13	Inujaponin D $C_{15}H_{22}O_4$ Mr 266 $(\alpha)_D^{23,1}$ - 14.67 (*c* 0.30, MeOH)	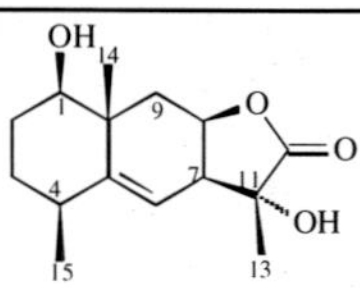	*I. japonica* (Wu, X.-D., et al., 2016)
14	Racemosalactone B $C_{15}H_{21}O_4Cl$ Mr 264.5	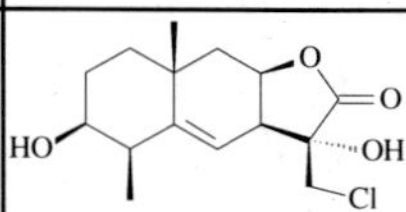	*I. rasemosa* (Ma, Y. Y., Zhao, D. G., Gao, K., 2013)
15	11β-hydroxy-13-chloro-eudesm-5-en-12,8-olide C15H21O3Cl Mr 284.5	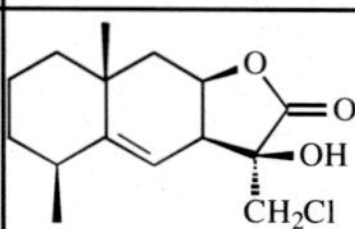	*I. helenium (Ding, Y., et al., 2019)*
16	Macrophyllilactone F C15H22O4 Mr 230	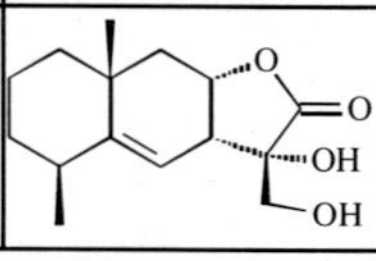	*I. macrophylla (Bo, Fu, et al., 2001)*

№	Sesquiterpene lactone	Structure	Sourse
17	1β,2β-Dihydroxyeudesma-4,11(13)-dien-12,8β-olide $C_{15}H_{20}O_4$ Mr 228	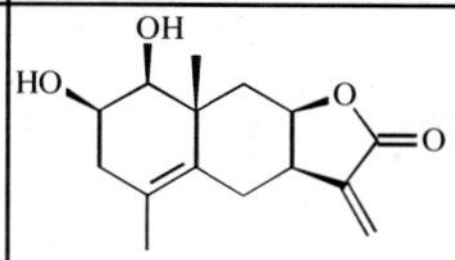	*I. japonica* (Qin, J. J., et al., 2010)
18	1β,3α-Dihydroxyeudesma-4,11(13)-dien-12,8β-olide $C_{15}H_{20}O_4$ Mr 228	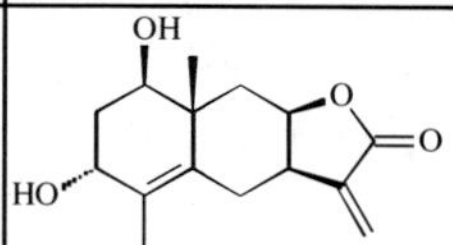	*I. japonica* (Qin, J. J., et al., 2010)
19	1β,3β-Dihydroxyeudesma-4,11(13)-dien-12,8β-olide $C_{15}H_{20}O_4$ Mr 228	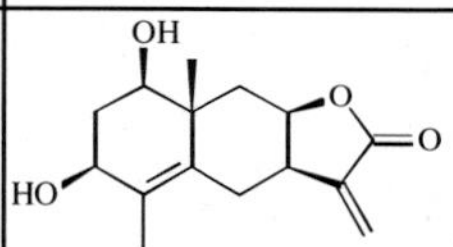	*I. japonica* (Qin, J. J., et al., 2010)
20	1β-Hydroxy-3-oxo-eudesma-4,11(13)-dien-12,8β-olide $C_{15}H_{18}O_4$ Mr 226	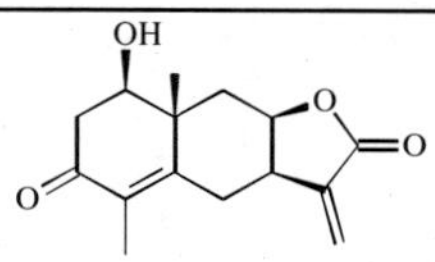	*I. japonica* (Qin, J. J., et al., 2010)
21	1β,3α-Dihydroxy-11αH-eudesm-4-en-12,8β-olide $C_{15}H_{22}O_4$ Mr 230	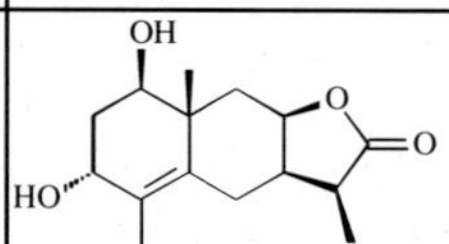	*I. japonica* (Qin, J. J., et al., 2010)
22	11α,13-Dihydro-isoalantolactone $C_{15}H_{22}O_2$ Mr 198	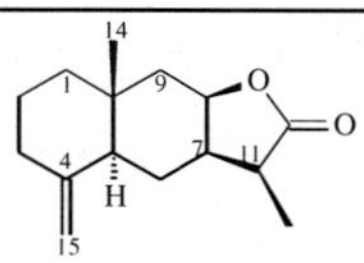	*I. racemosa* (Xu, L. W., Shi, Y. P., 2011)
23	1β,5α-Dihydroxy-11αH-eudesm-4(15)-en-12,8β-olide $C_{15}H_{22}O_4$ Mr 230	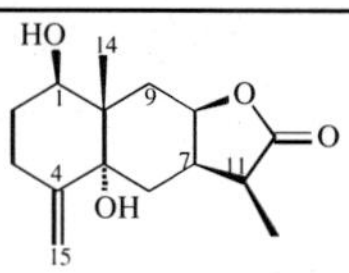	*I. japonica* (Qin, J.-J., et al., 2010)
24	Racemosalactone D C15H22O4 Mr 230	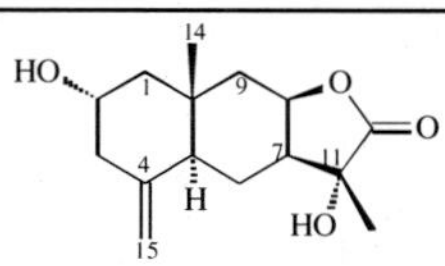	*I. rasemosa (Ma, Y. Y., Zhao, D. G., Gao, K., 2013)*

Table 6. (Continued)

№	Sesquiterpene lactone	Structure	Sourse
25	1β,6α-Dihydroxy-5αH-eudesma-3,11(13)-dien-12,8β-olide $C_{15}H_{20}O_4$ Mr 228 m.p. 155.7–158.2°C $(\alpha)_D^{20}$ - 21.5 (*c* 0.1, MeOH)	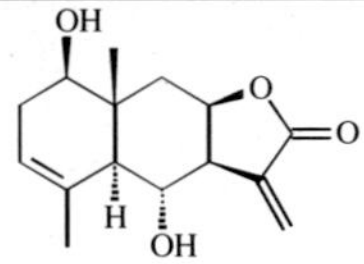	*I. britannica* (Zhang, X. F., et al., 2015)
26	1β,5α-Dihydroxy-4αH-eudesm-11(13)-ен-12,8β-olide $C_{15}H_{22}O_4$ Mr 230	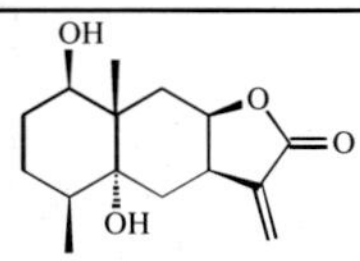	*I. japonica* (Qin, J. J., et al., 2010)
27	1β,5α-Dihydroxy-4αH, 11αH-eudesm-12,8β-olide $C_{15}H_{24}O_4$ Mr 290	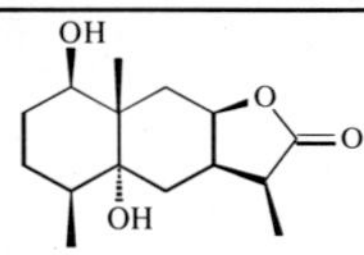	*I. japonica* (Qin, J. J., et al., 2010)
28	Inujaponin E $C_{15}H_{24}O_4$ Mr 268 $(\alpha)_D^{23,7}$-12.27 (*c* 0.28, MeOH)	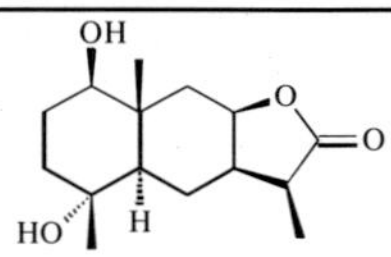	*I. japonica* (Wu, X. D., et al., 2016)
29	Racemosalactone A $C_{15}H_{22}O_3$ Mr 214	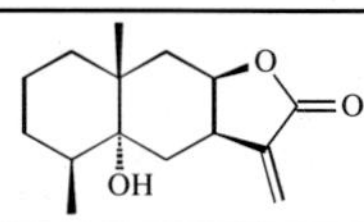	***I. rasemosa* (Ma, Y. Y., Zhao, D. G., Gao, K., 2013)**
30	Inujaponin G $C_{15}H_{20}O_4$ Mr 264 $(\alpha)_D^{26,5}$ + 60.00 (*c* 0.10, MeOH)	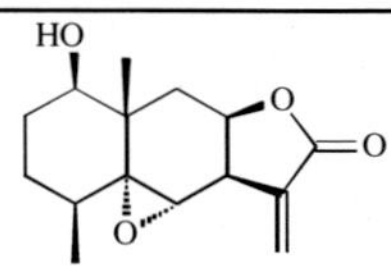	*I. japonica* (Wu, X. D., et al., 2016)
31	4α,15α-Epoxy-4α,15α-epoxypulchellin E $C_{17}H_{22}O_6$ Mr 322	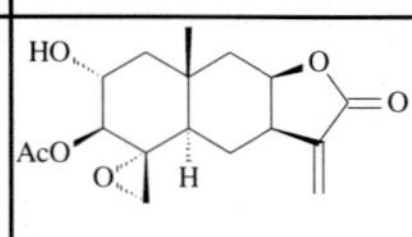	*I. oculus-christi* L. (Trendafilova, A., Ivanova, V., Todorova, M., Aneva, I., 2017)
32	Macrophyllilactone E $C_{15}H_{20}O_3$ Mr 212	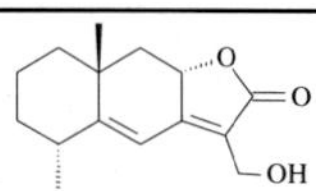	*I. racemosa* (Xu, L. W., Shi, Y. P., 2011) *I. helenium* (Qin, J. J., et al., 2010) *I. macrophylla* (Bo, Fu, et al., 2001)

№	Sesquiterpene lactone	Structure	Sourse
33	Macrophyllilactone G $C_{15}H_{18}O_6$ Mr 258	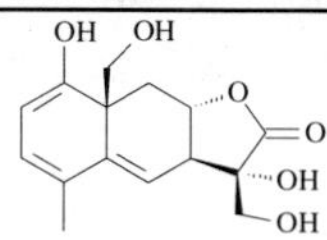	*I. macrophylla* (Bo, Fu, et al., 2001)
34	4β,5α-Dihydroxy-1-oxoeudesma-2,11(13)-dien-12,6β-olide $C_{15}H_{18}O_5$ Mr 242 $(\alpha)_D^{20}$ – 42.0 (*c* 0.025, MeOH)	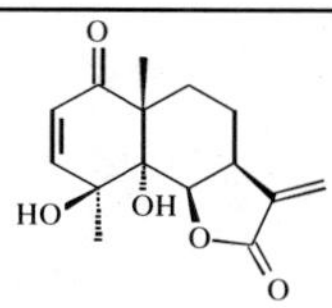	*I. wissmanniana* (Cheng, X. R., et al., 2013a)
35	3-Oxoeudesma-1,4,11(13)-trien-12,6β-olide $C_{15}H_{16}O_3$ Mr 208 m.p. 61–75°C $(\alpha)_D^{20}$ – 239.9 (*c* 0.10, MeOH)	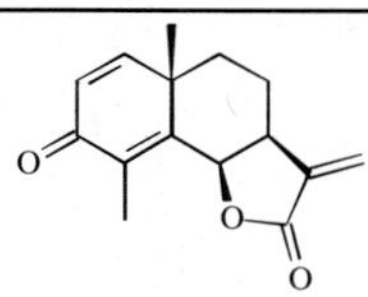	*I. wissmanniana* (Cheng, X. R., et al., 2013a)
36	Eudesma-3,11(13)-dien-12,5β-olide $C_{15}H_{20}O_2$ Mr 196 $(\alpha)_D^{20}$ – 14.7 (*c* 0.10, CH_2Cl_2)	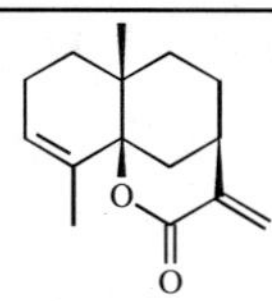	*I. wissmanniana* (Cheng, X. R., et al., 2013a)

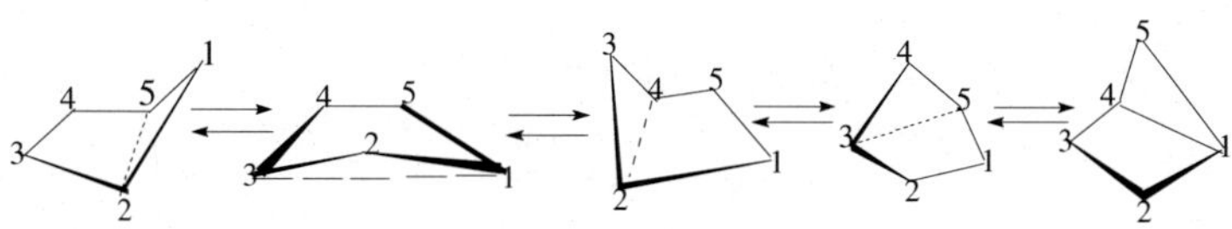

Figure 6. Types of conformations of the five-membered ring.

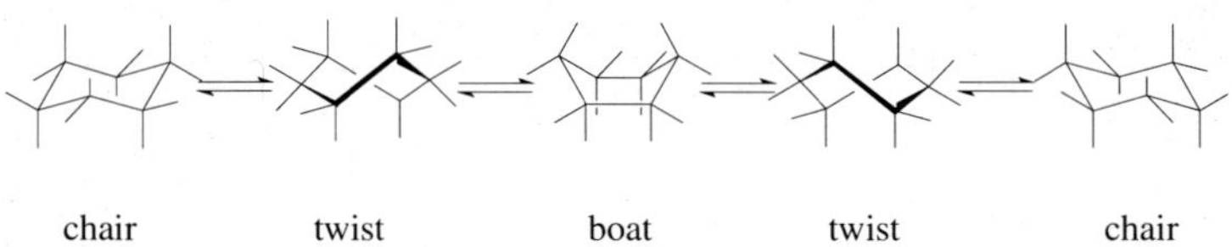

Figure 7. Types of conformations of six-membered rings.

The existing conformations of six-membered rings can be studied using the example of cyclohexane. Among several possible conformations of cyclohexane, the chair conformation has the lowest energy (i.e., more stable and preferable). There are also other conformations (Figure 7), such as boat conformation and twist conformation.

Table 7. Conformations of the eudesmanolide cycles of the *Inula* L. genus plants, established by X-ray diffraction analysis

№	Compound	Conformations	Literature
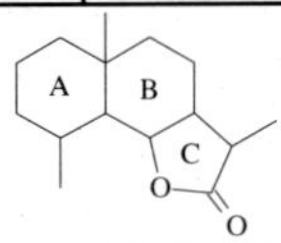			
1	Pulchellin C	cycle A has symmetrical chair conformation, cycle B has strongly distorted chair, cycle C has 7,8α-half-chair, respectively	(Adekenov, S. M., et al., 1990)
2	4(15)α-Epoxy-isoalantolactone	cycle A has chair conformation, cycle B has strongly distorted chair, cycle C has half-chair, respectively	(Kulyyasov, A. T., Seytembetov, T. S., Turdybekov, K. M., Adekenov, S. M., 1996)
3	4(15)α-Epoxy-pulchellin C	cycle A has chair conformation, cycle B has strongly distorted chair, cycle C has half-chair, respectively	(Tleugabulov, N. S., et al., 1998)
4	5α,6α-Epoxy-alantolacton	cycle A has chair conformation, cycle B has strongly distorted half-chair, cycle C has nearly flat conformation, respectively	(Tleugabulov, N. S., et al., 1998)
5	Gracilin	cycle A has distorted 1α,2β-half-chair conformation, cycle B has chair, cycle C has distorted half-chair, respectively	(Turmukhambetov, A. Zh., Adekenov, S. M., Turdybekov, K. M., Gatilov, Yu. V., 1991)
6	*Tourneforin*	cycle A has distorted half-chair conformation, cycle B has boat, *cis*–articulated lactone cycle C is flat, respectively	(Talzhanova, N. A., et al., 2007)
7	9β-Hydroxy-11β(H),13-dihydro-santamarine	cycle A has sofa conformation, cycle B has chair, cycle C has 7α-envelope, respectively	(Sham'yanov, I. D., Tashkhodzhaev, B., Mukhamatkhanova, R. F., Antipin, M. Yu., 2000a)
8	1α-Acetoxy-11(H),13-dihydroduglanin	cycle A has 10β-sofa conformation, cycle B has chair, cycle C has 7α-envelope, respectively	(Sham'yanov, I. D., Tashkhodzhaev, B., Mukhamatkhanova, R. F., Antipin, M. Yu., 2000b)
9	2-chloro-3-oxo-7α,6,11β(H)-eudesm-1,4-dien-6,12-olide	cycle A has flat conformation, cycle B has chair, cycle C has envelope, respectively	(Ivasenko, S. A., et al., 2006)

Table 7 shows the conformations of the eudesmanolide cycles of the *Inula* L. genus plants established by the X-ray diffraction method.

Thus, in most eudesmanolides, cycle A has the conformation of either a chair or a sofa; cycle B has either a chair or a distorted chair, and cycle C has an envelope or half-chair depending on the location of functional groups and double bonds.

Eudesmanes are also subdivided into three stereochemical series (Figure 8) (Semenov, A. A., 2000):

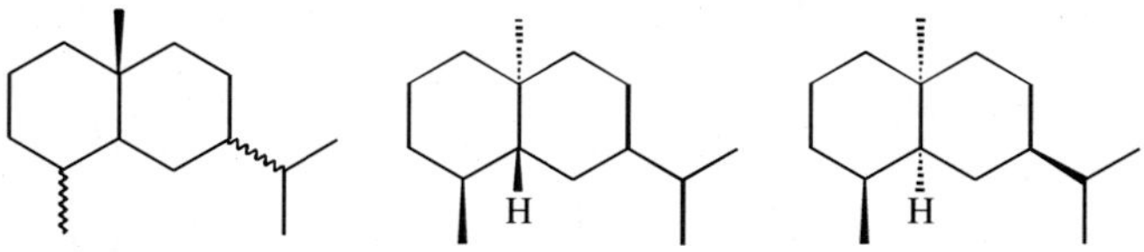

Eudesmanes Selinanes Steiractinanes

Figure 8. Stereochemical series of eudesmanolides.

Representatives of only eudesmanes (mainly) and selinanes are present in plants of the *Inula* L. genus.

Eudesmanolides are precursors of other series of sesquiterpene lactones: 1,10-*seco*- and 4,5-*seco*-eudesmanolides that are presented in Table 8.

Table 8. *Seco*-eudesmanolides of *Inula* L. genus plants

№	Sesquiterpene lactone	Structure	Source
1,10-*Seco*-eudesmanolides			
1	Britannilactone $C_{15}H_{22}O_4$ Mr 266	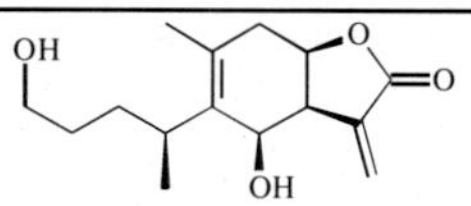	*I. japonica* (Xu, L. W., Shi, Y. P., 2011)
2	1-*O*-Acetyl-britannilactone $C_{17}H_{24}O_5$ Mr 308	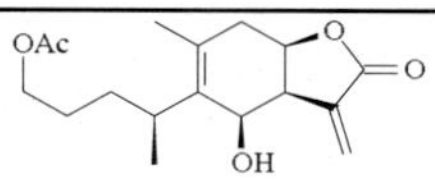	*I. britannica* (Hana, M., Wena, J. K., Zhenga, B., Zhang, D.-Q., 2004) *I. japonica* (Xu, L. W., Shi, Y. P., 2011)
3	1,6-Di-O-acetyl-britannilactone $C19H26O6$ Mr 350	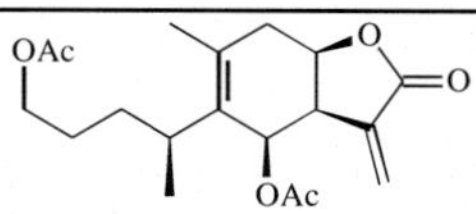	*I. japonica (Xu, L. W., Shi, Y. P., 2011)* *I. britannica (Chen, X., et al., 2017)*
4	1,6α-Dihydroxy-eriolanolide $C15H22O4$ Mr 266	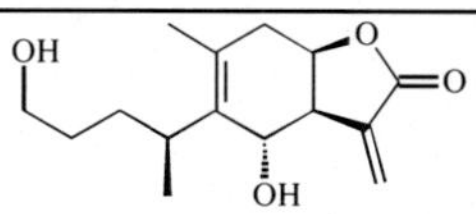	*I. japonica (Qin, J. J., et al., 2010)* *I. japonica (Chen, X., et al., 2015)*

Table 8. (Continued)

№	Sesquiterpene lactone	Structure	Source
5	1-Acetoxy-6α-hydroxy- eriolanolide $C_{17}H_{24}O_5$ Mr 308	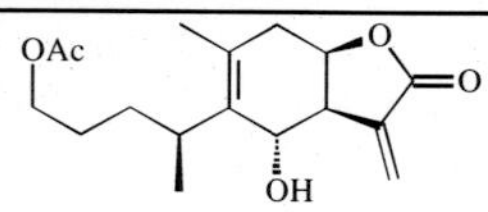	*I. japonica* (Qin, J. J., et al., 2010) *I. britannica* (Zhang, X. F., et al., 2015)
6	6α-Acetyloxy-1-hydroxy-4αH-1,10-*seco*-eudesma-5(10),11(13)-dien-12,8β-olide $C_{17}H_{24}O_5$ Mr 308	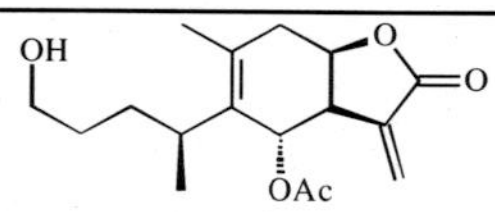	*I. japonica* (Qin, J. J., et al., 2010)
7	(4*S*,6*S*,7*S*,8*S*,9*S*)-1-Acetoxy-6,9-dihydroxy-1,10-*seco*-eudesma-5(10), 11(13)-dien-12,8-olide $C_{17}H_{24}O_6$ Mr 324 m.p. 88.1–90.3°C $(\alpha)_D^{20}$ +66.4	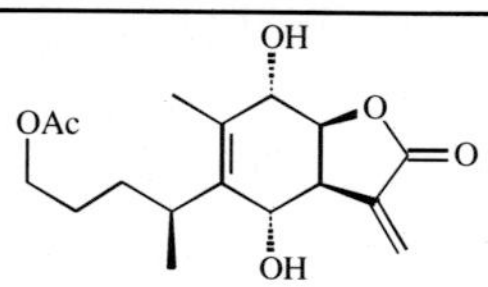	*I. britannica* (Zhang, X. F., et al., 2015)
8	1,10β-Dihydroxy-4αH-1,10-*seco*-eudesma-5, 11(13)-dien-12,8β-olide $C_{15}H_{22}O_4$ Mr 230	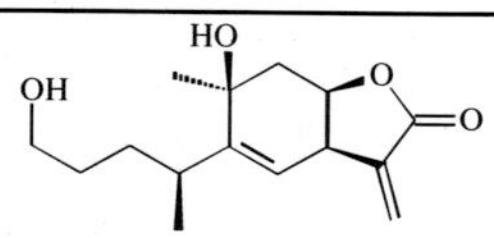	*I. japonica* (Qin, J. J., et al., 2010)
9	1,14-Dihydroxy-4αH-1,10- *seco*-eudesma-5(10),11(13)-dien-12,8β-olide $C_{15}H_{22}O_4$ Mr 230	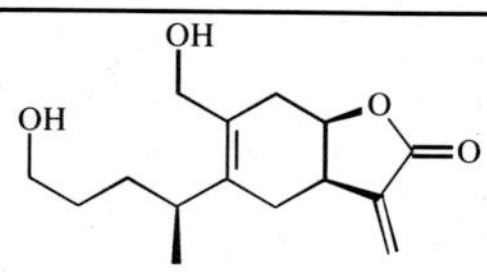	*I. japonica* (Qin, J. J., et al., 2010)
10	1-Acetyloxy-6α,14-dihydroxy-4αH-1,10-*seco*-eudesma-5(10), 11(13)-dien-12,8β-olide $C_{17}H_{24}O_6$ Mr 324	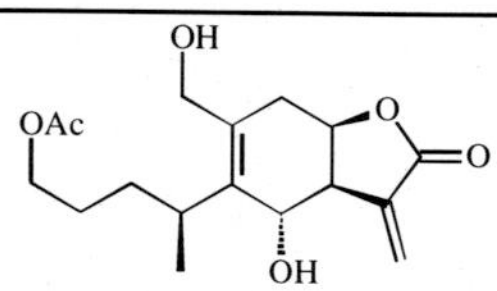	*I. japonica* (Qin, J. J., et al., 2010)
11	1-Hydroxy-6β-iso-butyryloxy-4αH-1,10-seco-eudesma-5(10), 11(13)-dien-12,8β-olide C19H28O5 Mr 336	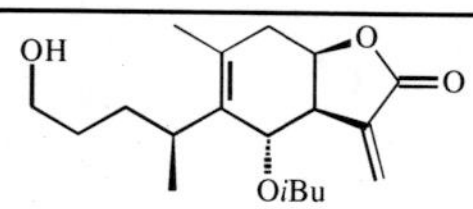	*I. japonica (Qin, J. J., et al., 2010; Chen, X., et al., 2015)*

№	Sesquiterpene lactone	Structure	Source
12	14-Isobutyloxy-6-deoxy-britannilactone $C_{19}H_{28}O_5$ Mr 336 $(a)_D^{20}$ + 122.9 (*c* 0.07, $CHCl_3$)	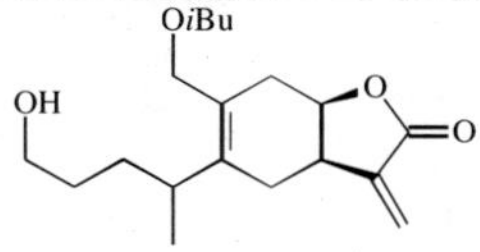	*I. britannica* (Fischedick, J. T., et al., 2013)
13	1-Hydroxy-6α-(2-methyl-butyryloxy)-4αH-1,10-*seco*-eudesma-5(10), 11(13)-dien-12,8β-olide $C_{20}H_{30}O_5$ Mr 350	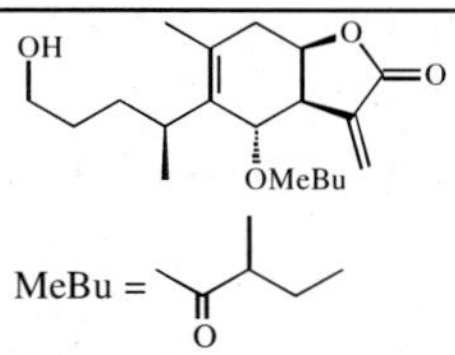	*I. japonica* (Qin, J. J., et al., 2010)
14	14-(3-Methylbutanoyl)-6-deoxybritannilactone $C_{20}H_{30}O_5$ Mr 350	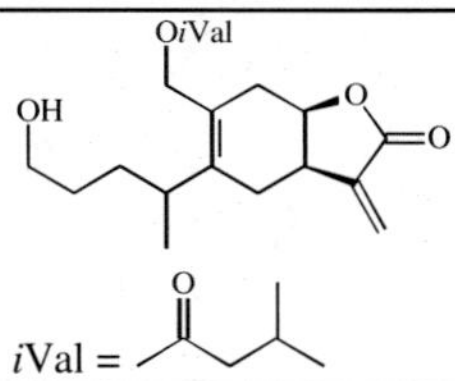	*I. britannica* (Fischedick, J. T., et al., 2013)
15	1-Hydroxy-6α-(isovaleryl-oxy)-4αH-1,10-*seco*-eudesma-5(10),11(13)-dien-12,8β-olide $C_{20}H_{30}O_5$ Mr 350	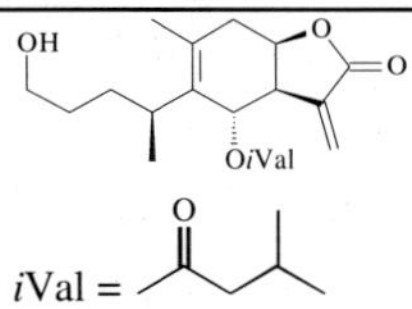	*I. japonica* (Qin, J. J., et al., 2010; Chen, X., et al., 2015)
16	1-Hydroxy-6α-((3-methylvaleryl)oxy)-4αH-1,10-*seco*eudesma-5(10),11(13)-dien-12,8β-olide $C_{21}H_{32}O_5$ Mr 364 $(\alpha)_D^{20}$ + 109.5 (*c* 0.15, $CHCl_3$)	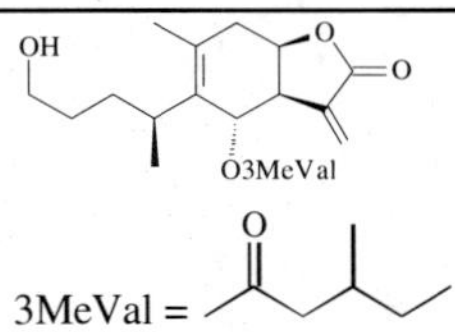	*I. japonica* (Qin, J. J., et al., 2010) *I. britannica* (Fischedick, J. T., et al., 2013)
	4,5-*Seco*-eudesmanolides		
1	Umbellifolide $C_{15}H_{20}O_4$ Mr 228	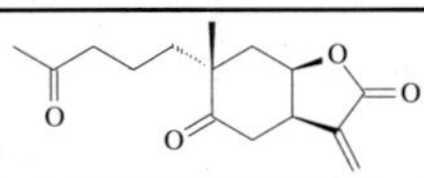	*I. helenium* (Jiang, H. L., et al., 2011)
2	(5*R*,7*R*,10*S*)-4,5-Epoxy-4,5-*seco*-eudesma-1,3-dien-12,5-olide $C_{15}H_{18}O_3$ Mr 210 $(\alpha)_D^{20}$ – 211.8 (*c* 0.10, MeOH)	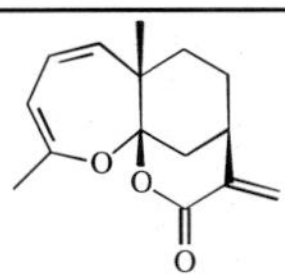	*I. wissmanniana* (Cheng, X. R., et al., 2013a)

Guaianolides and Pseudoguaianolides

Guaianolides belong to the large groups of sesquiterpene lactones which skeleton consists of fused five- and seven-membered rings and a lactone cycle. The structure of pseudoguaianolides differs from guaianolides by the location of the C15 methyl group: in guaianolides it is located at C4, while in pseudoguaianolides is at C5 respectively.

Lactones of these groups can also be linear (8,12-olides) and non-linear (6,12-olides). Table 9 shows the guaianolides and pseudoguaianolides of the *Inula* L. genus plants.

Table 9. Guaianolides and pseudoguaianolides of *Inula* L. genus plants

№	Sesquiterpene lactone	Structure	Source
Guaianolides			
1	Inuviscolide $C_{15}H_{20}O_3$ Mr 248	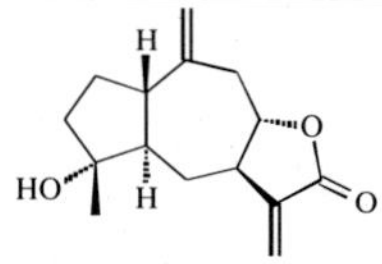	*I. racemosa* (Xu, L. W., Shi, Y. P., 2011)
2	2α-Acetoxy-4α-hydroxy-1βH-guaia-10(14),11(13)-dien-12,8α-olide $C_{17}H_{22}O_5$ Mr 306	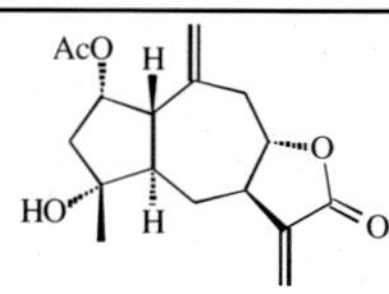	*I. hupehensis* (Ren, J., et al., 2013)
3	2α-Acetoxy-4α,9β-dihydroxy-1βH-guaia-11(13),10(14)-dien-12,8α-olide C17H22O6 Mr 322	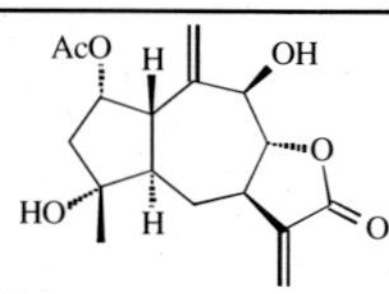	*I. oculus-christi (Trendafilova, A., Ivanova, V., Todorova, M., Aneva, I., 2017)*
4	8-Epiinuviscolide C15H20O3 Mr 212	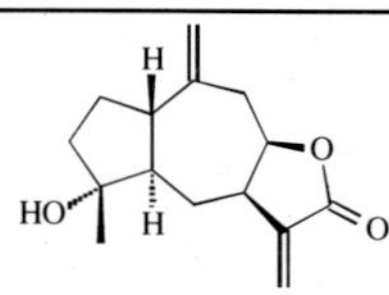	*I. helianthus-aquatilis (Hua, Y., et al., 2012)*
5	Gaillardin C17H22O5 Mr 306 m.p. 199-200°C (α)D20 -15°	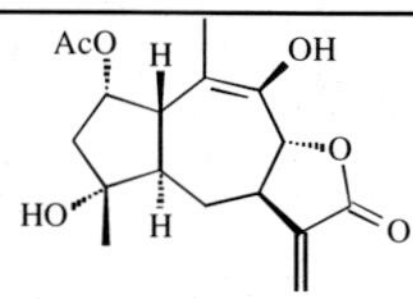	*I. oculus-christi (Aljancic, I. S., et al., 2008)*

№	Sesquiterpene lactone	Structure	Source
6	2α-Acetoxy-4α,6α-dihydroxy-1βH-guaia-9,11(13)-dien-12,8α-olide $C_{17}H_{22}O_6$ Mr 322	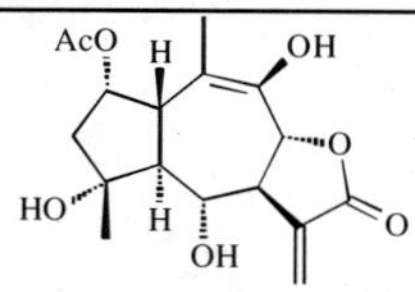	*I. lineariifolia* (Qin, J. J., et al., 2013)
7	Hupehenolide M $C_{17}H_{22}O_7$ Mr 338 $(\alpha)_D^{20}$ + 69.3 (*c* 0.10, MeOH)	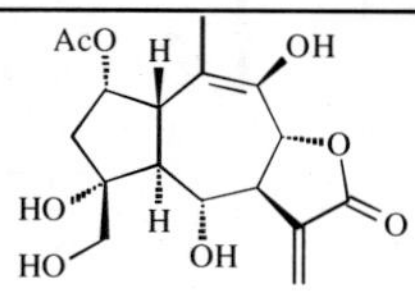	*I. hupehensis* (Ren, J., et al., 2013)
8	Inuchinenolide B $C_{17}H_{22}O_5$ Mr 306	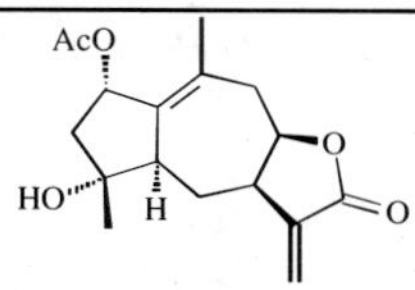	*I. lineariifolia* (Qin, J. J., et al., 2013)
9	Inujaponin H $C_{15}H_{22}O_4$ Mr 266 $(\alpha)_D^{23,5}$ -39.27 (*c* 0.06, MeOH)	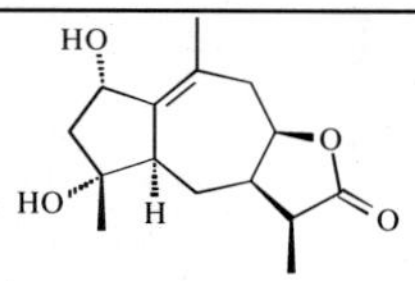	*I. japonica* (Wu, X. D., et al., 2016)
10	11,13-Dihydro-inuchinenolide B $C_{17}H_{22}O_5$ Mr 306 $(a)_D^{20}$ - 46.1 (*c* 0.26, $CHCl_3$)	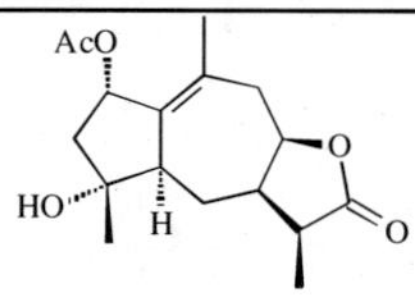	*I. britannica* (Fischedick, J. T., et al., 2013)
11	Deacetyl-inuchinenolide B C17H22O6 Mr 322	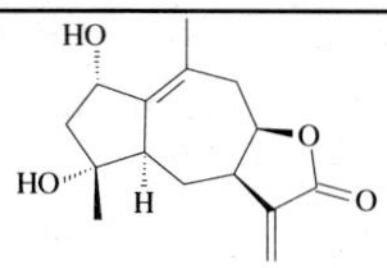	*I. lineariifolia (Fischedick, J. T., et al., 2013)*
12	2α-Acetoxy-5α,6α-epoxy-4α-hydroxy-1β,7αH-guaia-11(13)-en-12,8α-olide C17H22O6 Mr 322	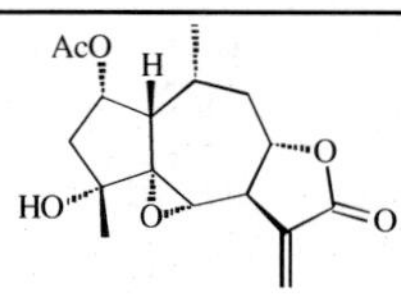	*I. hupehensis (Ren, J., et al., 2013)*

Table 9. (Continued)

№	Sesquiterpene lactone	Structure	Source
13	9α,10α-Epoxy-2-epi-gaillardin $C_{17}H_{22}O_6$ Mr 322	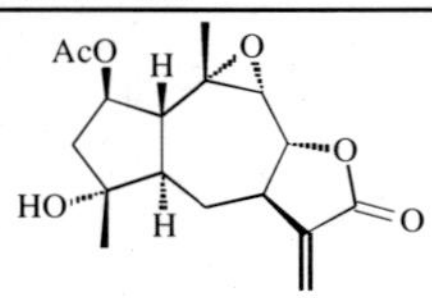	*I.oculus-christi* (Trendafilova, A., Ivanova, V., Todorova, M., Aneva, I., 2017)
14	9β,10β-epoxygaillardin $C_{17}H_{22}O_6$ Mr 322	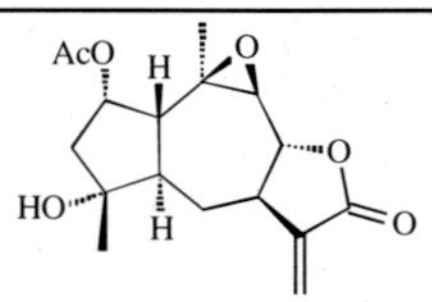	*I. oculus-christi* (Trendafilova, A., Ivanova, V., Todorova, M., Aneva, I., 2017)
15	Dehydrocostuslactone $C_{15}H_{18}O_2$ Mr 194	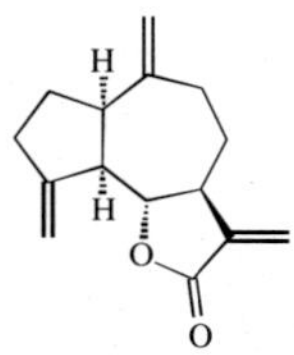	*I. helenium* (Wu, X. D., et al., 2008)
16	Inulonolide $C_{15}H_{18}O_4$ Mr 262	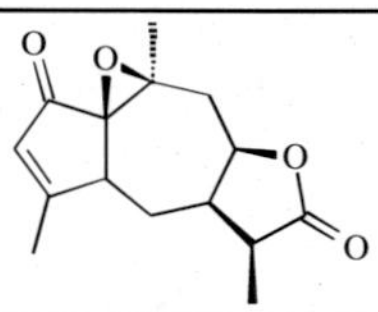	*I. britannica* (Dusmatova, D. E., et al., 2020)
Pseudoguaianolides			
1	2α,4β-Dihydroxy-pseudoguai-11(13)-en-12,8α-olide $C_{15}H_{22}O_4$ Mr 230	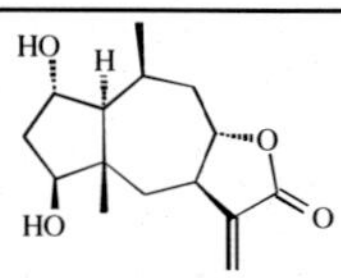	*I. japonica* (Qin, J. J., et al., 2010)
2	2α-Acetyloxy-4β-hydroxy-pseudoguai-11(13)-en-12,8α-olide C17H24O5 Mr 308	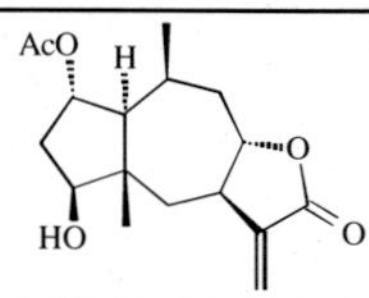	*I. japonica (Qin, J. J., et al., 2010)*
3	Neohupehenolide A C15H20O4 Mr 228 (α)D20 + 230.9 (c 0.04, CH2Cl2)	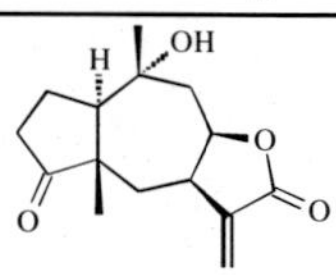	*I. hupehensis (Ren, J., et al., 2013)*

№	Sesquiterpene lactone	Structure	Source
4	Neohupehenolide B $C_{15}H_{20}O_6$ Mr 260 $(\alpha)_D^{20}$ + 142.6 (*c* 0.10, MeOH)	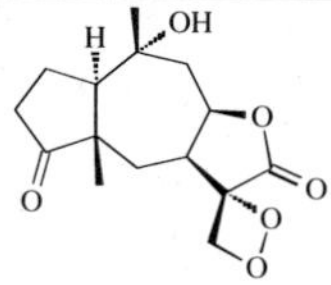	*I. hupehensis* (Ren, J., et al., 2013)
5	2-Desoxy-4-epipulchellin $C_{15}H_{22}O_3$ Mr 214	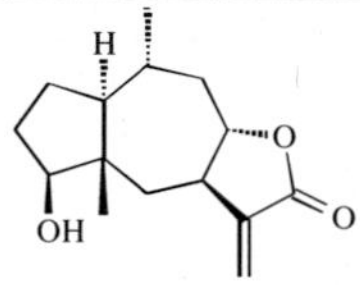	*I. britannica* (Yang, J. L., Wang, R., Liu, L. L., Shi, Y. P., 2011) *I. helianthus-aquatilis* (Hua, Y., et al., 2012)
6	4β-Acetoxy-6α-hydroxy-10βH-pseudoguai-11(13)-en-12,8α-olide $C_{17}H_{24}O_5$ Mr 308	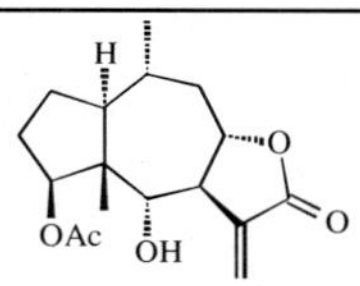	*I. helianthus-aquatilis* (Hua, Y., et al., 2012)
7	2-*O*-Acetyl-4-epi-pulchellin $C_{17}H_{24}O_5$ Mr 308	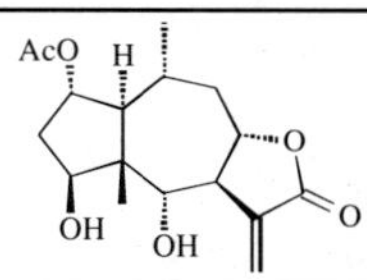	*I. helianthus-aquatilis* (Hua, Y., et al., 2012)
8	6α-Acetoxy-4β-hydroxy-10βH-pseudoguai-11(13)-en-12,8α-olide $C_{17}H_{24}O_5$ Mr 308	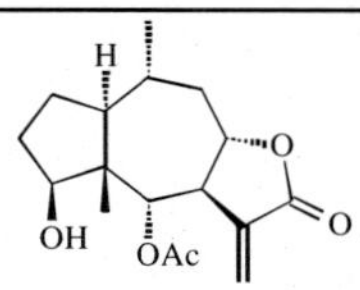	*I. helianthus-aquatilis* (Hua, Y., et al., 2012)
9	Britanin $C_{19}H_{26}O_7$ Mr 366 m.p. 192-194°C $(\alpha)_D^{20}$ -26°	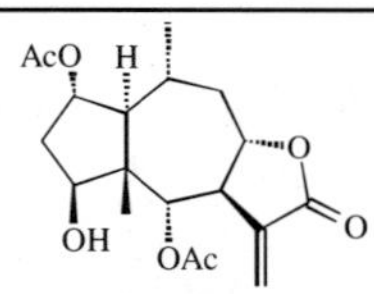	*I. britannica* (Dusmatova, D. E., et al., 2018)
10	Hupehenolide E C15H20O4 Mr 228 (α)D20 + 100.0 (c 0.40, CH2Cl2)	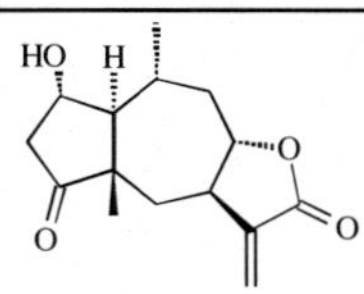	*I. hupehensis (Ren, J., et al., 2013)*
11	Ergolide C17H22O5 Mr 306	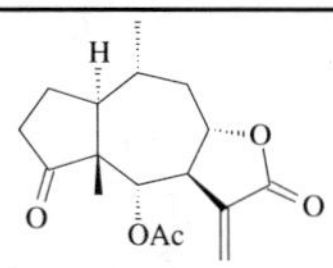	*I. britannica (Chun, J. K., et al., 2007)*

Table 9. (Continued)

№	Sesquiterpene lactone	Structure	Source
12	(1*S*,2*R*,5*S*,7*R*,8*S*,10*R*)-2-Methoxy-4-oxy-pseudoguai-11(13)-en-12,8-olide $C_{16}H_{22}O_4$ Mr 278 $(\alpha)_D^{27}$ + 70.1 (*c* 0.10, MeOH)	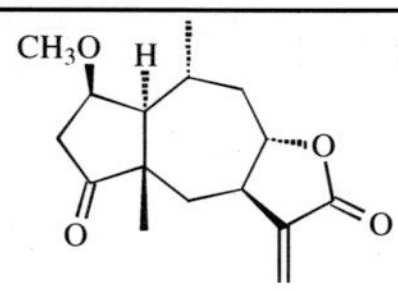	*I. japonica* (Yu, Z. P., et al., 2019)
13	11β,13-Dihydroergolide $C_{17}H_{22}O_5$ Mr 306	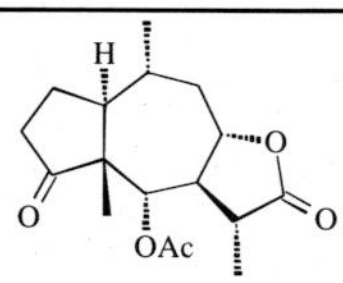	*I. helianthus-aquatilis* (Hua, Y., et al., 2012)
14	Hupehenolide B $C_{15}H_{20}O_4$ Mr 228 $(\alpha)_D^{20}$ + 44.4 (*c* 0.10, CH_2Cl_2)	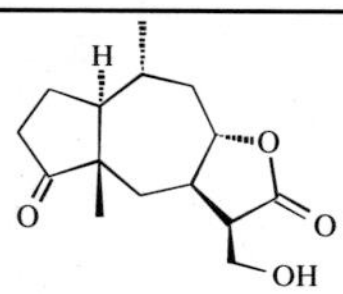	*I. hupehensis* (Ren, J., et al., 2013)
15	Aromaticin $C_{15}H_{18}O_3$ Mr 246 m.p. 232-234°C $(\alpha)_D$ -6° $(\alpha)_D^{27}$ – 42.8 (*c* 0.29, MeOH)	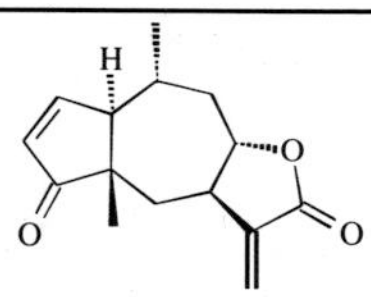	*I. japonica* (Yu, Z. P., et al., 2019)
16	Hookerolide A $C_{19}H_{26}O_6$ Mr 350	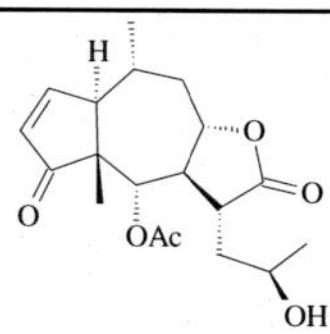	*I. hookeri* (Cheng, X. R., et al., 2013b)
17	HOOKEROLIDE B C19H26O6 MR 350	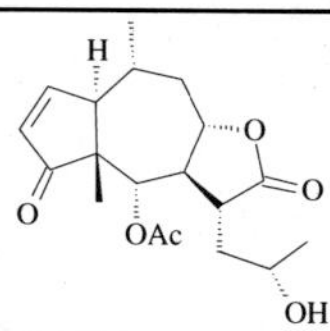	*I. hookeri (Cheng, X. R., et al., 2013b)*
18	HOOKEROLIDE C C19H26O7 MR 366	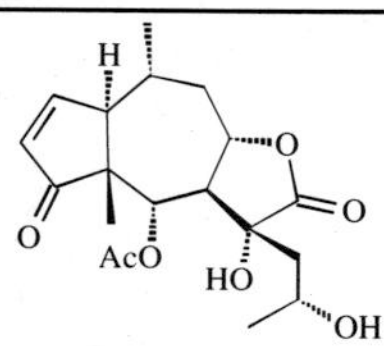	*I. hookeri (Cheng, X. R., et al., 2013b)*

№	Sesquiterpene lactone	Structure	Source
19	Hookerolide D $C_{19}H_{26}O_7$ Mr 366	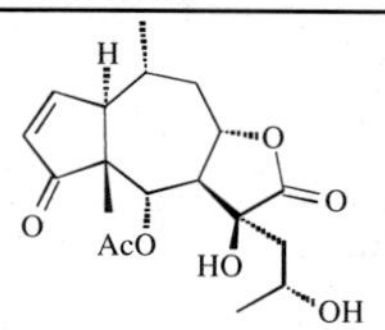	*I. hookeri* (Cheng, X. R., et al., 2013b)

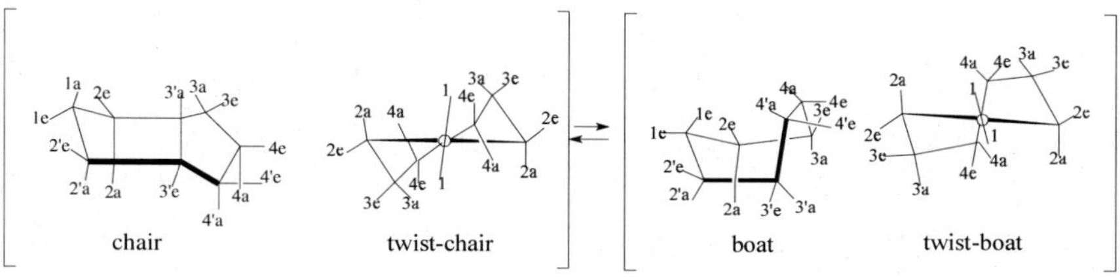

Figure 9. Cycloheptane conformations.

The conformations of guaianolides and pseudoguaianolides are similar. The five-membered ring conformations have been described above (refer to the Eudesmanolides section). The seven-membered ring conformation can be illustrated by the example of cycloheptane. Conformational changes for seven-membered rings can be determined using physical and computational methods: NMR spectroscopy, vibrational spectroscopy, electron diffraction, force field calculations and X-ray structural analysis of (crystalline) substances. The conformations of cycloheptane are represented by two families shown in Figure 9 (Eliel, E. L., Wilen, S. H., 1994).

The first one is formed by the chair and twist-chair conformations, while the second one is formed by the boat and twist-boat conformations. The conformations of cycloheptane differ from that of cyclohexane by overlap. Therefore, the twist-chair and twist-boat conformations are the most stable for cycloheptane in the process of pseudorotation. In cyclohexane, the chair is almost free of tention, while in cycloheptane even the most stable twist chair cannot completely avoid the occlusion interactions. The families of chair-twist-chair and boat-twist-boat in case of cycloheptane can transform into each other by only overcoming a relatively high barrier of 35.6 kJ/mol (Eliel, E. L., Wilen, S. H., 1994). Unsaturated cycloheptene mainly attains the shape of a chair as the C = C fragment takes the place of one cyclic carbon atom in the chair of cyclohexane (https://studref.com/532205/matematika_himiya_fizik/tsikly_bolshego_razmera).

Table 10 shows the conformations of the cycles of natural guaianolides and pseudoguaianolides, determined by X-ray diffraction analysis.

Table 10. Ring conformations of natural guaianolides and pseudoguaianolides determined by X-ray diffraction analysis

№	Sesquiterpene lactone	Structure	Source
Guaianolides 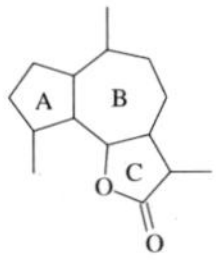			
1	Anolide	cycle A has envelope conformation, cycle B has chair, cycle C has strongly distorted 7α-envelope conformation, respectively	(Turdybekov, K. M., Turmukhambetov, A. Zh., Adekenov, S. M., Struchkov, Yu. T, 1994)
2	1β,10α-Dioxyarglabine	cycle A has envelope confor-mation, cycle B has twist-chair, cycle C has half-chair conformation, respectively	(Adekenov, S. M., et al., 1993)
3	Isoepoxyestafiatin	cycle A has 1β- envelope conformation, cycle B has chair, cycle C has half-chair conformation, respectively	(Ponomarev, D. L., et al., 1997)
4	8-epi-desacyl-cinaropicrine	cycle A has distorted envelope conformation, cycle B has distorted chair, cycle C has slightly distorted envelope conformation, respectively	(Adekenov, S. M., Kadirberlina, G. M., Turdybekov, K. M., Struchkov, Yu. T., 1991b)
5	Isolipidiol	cycle A has distorted 3α-envelope conformation, cycle B has 1,5α,8β-chair, cycle C has half-chair conformation, respectively	(Fazylova, A. S., et al., 2000)
6	Tetrahydroleucomisin	cycle A has half-chair conformation, cycle B has chair, cycle C has envelope conformation, respectively	(Makhmudov, M. K., Plutno, A. B., Sham'yanov, I. D., Tashkhodzhaev, B., 1995)
7	Crepidioside A	cycle A has flat form, cycle B has 7α,1,10β-chair conforma-tion, cycle C has half-chair, respectively	(Fazylova, A. S., et al., 2000)
Pseudoguaianolides 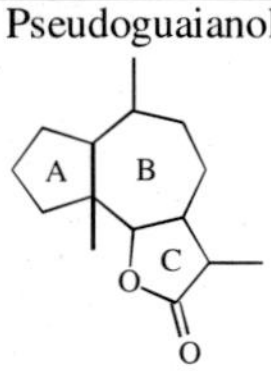			

№	Sesquiterpene lactone	Structure	Source
1	Inuchinenolide C	cycle A has between the envelope and the half-chair conformation, cycle B has between the chair and the twist-chair, cycle C has between the half-chair and the envelope conformation	(Adekenov, S. M., 1990b)
2	Erigerolide	cycle A has half-chair conformation, cycle B has twist-chair, cycle C has half-chair conformation, respectively	(Makhmudov, M. K., Tashkhodzhaev, B., Zhonkhodzhaeva, F. B., Sham'yanov, I. D., 1993b)

Based on the data analysis of Table 10, in the majority of guaianolides the five-membered cycle A takes the form of an envelope, the seven-membered cycle B has a chair or a distorted chair conformation; in pseudoguaianolides cycle A has half-chairs, cycle B has twist-chairs conformations, respectively.

Guaianolides are the precursors of another series of sesquiterpene lactones - 4,5-*seco*-guaianolides or xanthanolides found in plants of the *Inula* L. genus. Table 11 shows the representatives of the xanthanolides.

Table 11. Xanthanolides (4,5-*seco*-guaianolides) of *Inula* L. genus plants

№	Sesquiterpene lactone	Structure	Source
1	Tomentosin $C_{15}H_{20}O_3$ Mr 248	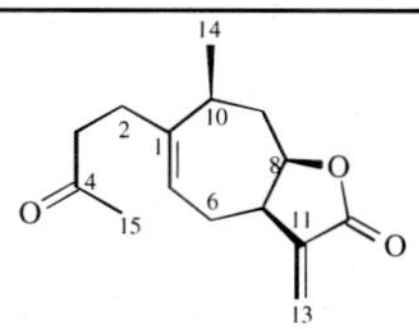	*I. japonica* (Qin, J. J., et al., 2010) *I. candida* subsp. *candida and* subsp. *decalvans* (Michalakea, E., et al., 2019)
2	8-Epitomentosin $C_{15}H_{20}O_3$ Mr 212	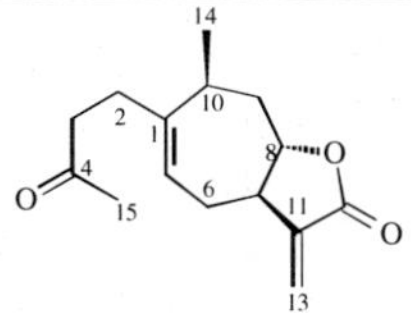	*I. helenium* (Vasas, A., Hohmann, J., 2011)
3	Britanlin G $C_{17}H_{22}O_5$ Mr 306	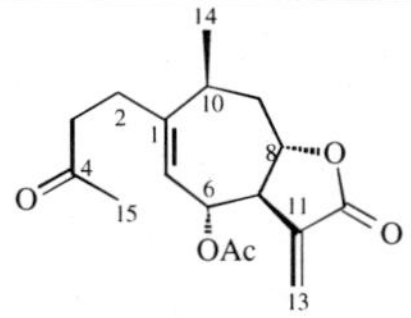	*I. leariifolia* (Qin, J. J., et al., 2013)

Table 11. (Continued)

№	Sesquiterpene lactone	Structure	Source
4	Inujaponin I $C_{20}H_{28}O_5$ Mr 348 $(\alpha)_D^{23,3}$- 68.33 (*c* 0.29, MeOH)	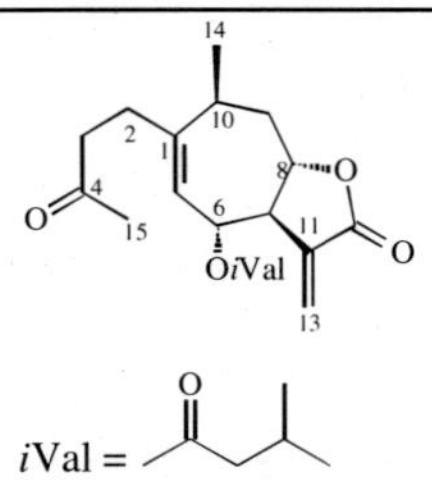	*I. japonica* (Wu, X. D., et al., 2016)
5	11α,13-Dihydro-tomentosin $C_{15}H_{22}O_3$ Mr 214	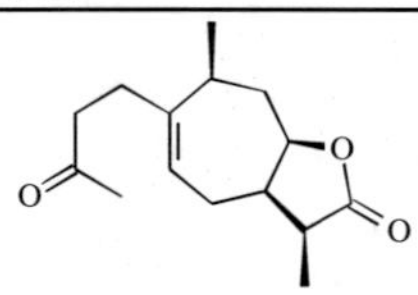	*I. hupehensis* (Chun, J. K., et al., 2007)
6	8-Epixanthatin-1β,5β-epoxide $C_{15}H_{18}O_3$ Mr 210	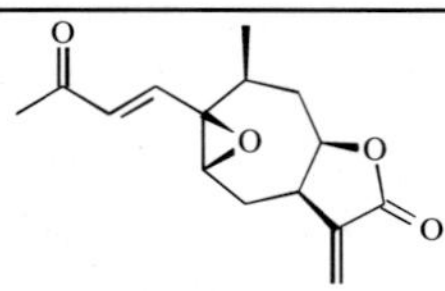	*I. hupehensis (Qin J. J., et al., 2012)*
7	Sundiversifolide $C_{14}H_{22}O_3$ Mr 238	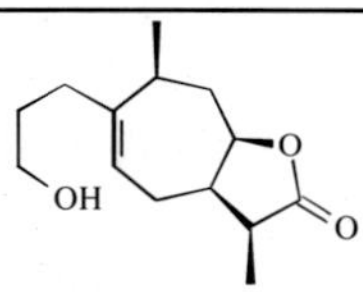	*I. hupehensis* (Qin J. J., et al., 2012)
8	11-Episundiversifolide $C_{14}H_{22}O_3$ Mr 238	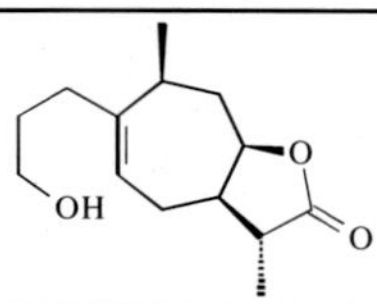	*I. hupehensis* (Qin J. J., et al., 2012)
9	Ivalbatin $C_{15}H_{20}O_4$ Mr 228	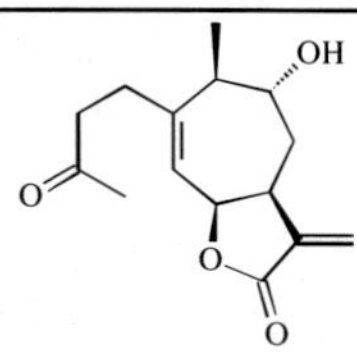	*I. viscosa, I. helenium, I. royleana* (Vasas, A., Hohmann, J., 2011)

Elemanolides

Elemanolides can be consideredas 2,3-*seco*-eudesmanolides. Table 12 shows elemanolides of plants of the *Inula* L. genus.

Eremophilanolides

Eremophylanolides differ from eudesmanolides by the location of the C14 methyl group: in eudesmanolides, it is at C10 while in eremophylanolides it is at C5 respectively. Most eremophylanolides are unsaturated and linear in structure. In Figure 10 shows the elemanolides Racemosalactone E from *I. rasemosa* (Ma, Y. Y., Zhao, D. G., Gao, K., 2013) and Inuloxine C from *I. viscosa* (Andol fi, A., et al., 2013).

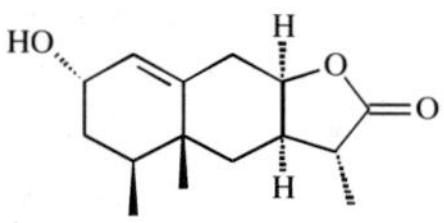

Racemosalactone E

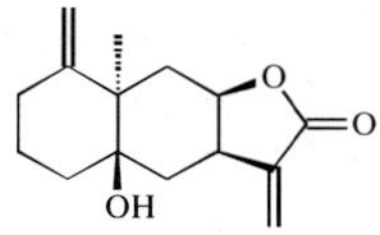

Inuloxine C

Figure 10. Examples of elemanolides of plants of the genus *Inula* L.

Table 12. Elemanolides of the *Inula* L. genus plants

№	Sesquiterpene lactone	Structure	Source
1	Igalane $C_{15}H_{20}O_2$ Mr 196 m.p. 79-79,5°C $(\alpha)_D^{20}$ +105°	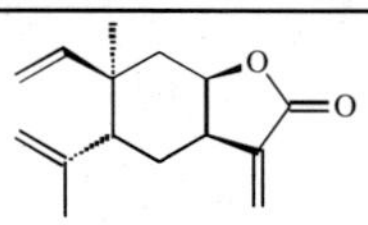	*I. helenium* (Nikonova, L. P., Nikonov, G. K., 1970) *I. grandis* (Dusmatova, D. E., et al., 2019)
2	Macrophyllilactone A $C_{15}H_{20}O_3$ Mr 212	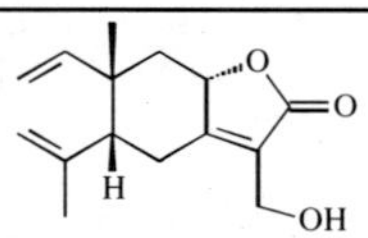	*I. macrophylla* (Bo, Fu, et al., 2001)
3	Macrophyllilactone C $C_{15}H_{22}O_4$ Mr 230	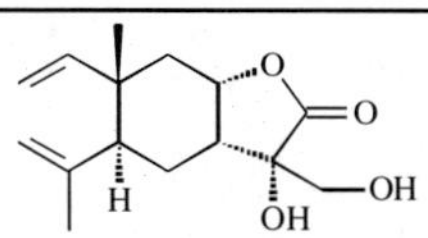	*I. macrophylla* (Bo, Fu, et al., 2001)
4	Macrophyllilactone B $C_{15}H_{22}O_4$ Mr 230	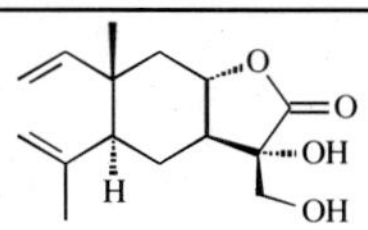	*I. macrophylla* (Bo, Fu, et al., 2001)
5	Macrophyllilactone D $C_{15}H_{22}O_4$ Mr 230	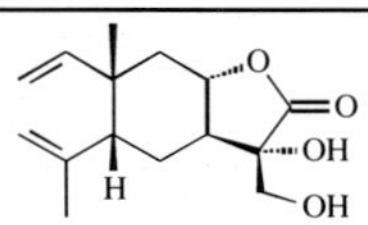	*I. macrophylla* (Bo, Fu, et al., 2001)

Other Types of *Inula* L. Genus Sesquiterpene Lactones

Table 13 demonstrates other types of sesquiterpene lactones of the *Inula* L. genus plants.

Table 13. Other types of sesquiterpene lactones of the *Inula* L. genus plants

№	Sesquiterpene lactone	Structure	Source
1	Carabrone $C_{15}H_{20}O_3$ Mr 248	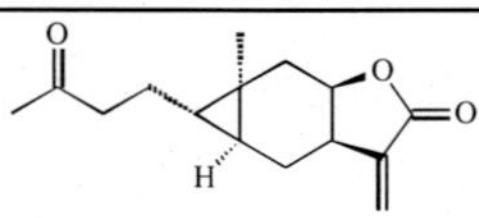	*I. cappa* (Wang, F. Y., et al., 2012) *I. helenium* (Vasas, A., Hohmann, J., 2011) *I. pterocaula* (Zhu, Sh., et al., 2019)
2	Inuloxine D $C_{15}H_{22}O_3$ Mr 214 $(\alpha)_D^{25}$ +81,8 (*c* 0.2)	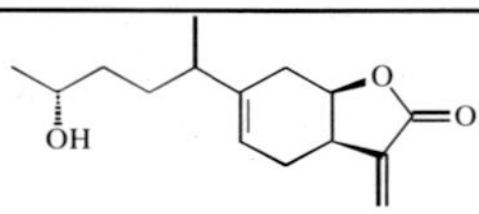	*I. viscosa* (Andolfi, A., et al., 2013)
3	(3a*S*,8a*R*)-6-methyl-3-methylidene-7-(3-oxobutyl)-3a,4,5,8a-tetrahydro-2H-cyclohepta(b)furan-2,8(3H)dione $C_{15}H_{18}O_4$ Mr 262	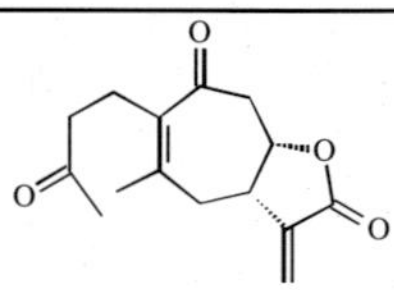	*I. candida* subsp. *candida* (Michalakea, E., et al., 2019) *I. candida* subsp. *decalvans* (Michalakea, E., et al., 2019)
4	Inusoniolide $C_{15}H_{21}O_3$ Mr 249	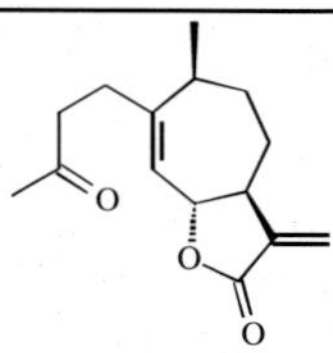	*I. candida* subsp. *candida* (Michalakea, E., et al., 2019) *I. candida* subsp. *decalvans* (Michalakea, E., et al., 2019) *I. verbascifolia* (Michalakea, E., et al., 2019)
5	4-*O*-Dihydro-inusoniolide $C_{14}H_{22}O_3$ Mr 238	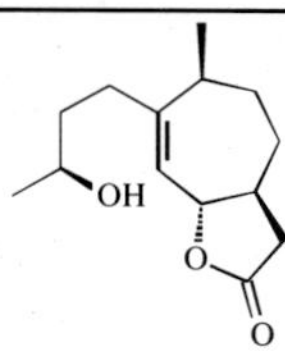	*I. verbascifolia* (Michalakea, E., et al., 2019)
6	Britanlin H $C_{19}H_{26}O_7$ Mr 366	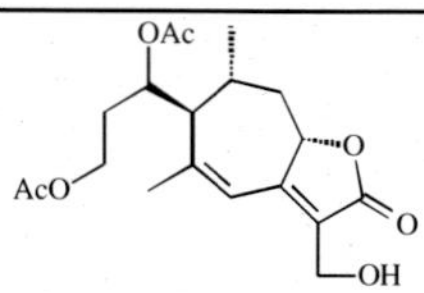	*I. britannica* (Yang, J. L., Wang, R., Liu, L. L., Shi, Y. P., 2011)

Dimeric and Trimeric Sesquiterpene Lactones

Plants of the genus *Inula* L. also produce dimeric sesquiterpene lactones, which monomers can be lactones of the same or different types of skeleton (Table 14).

Table 14. Dimeric and trimeric sesquiterpene lactones of the *Inula* L. genus plants

№	Sesquiterpene lactone	Structure	Source
Dimeric sesquiterpene lactones			
1	Macrophyllidimer A $C_{30}H_{40}O_4$ Mr 464	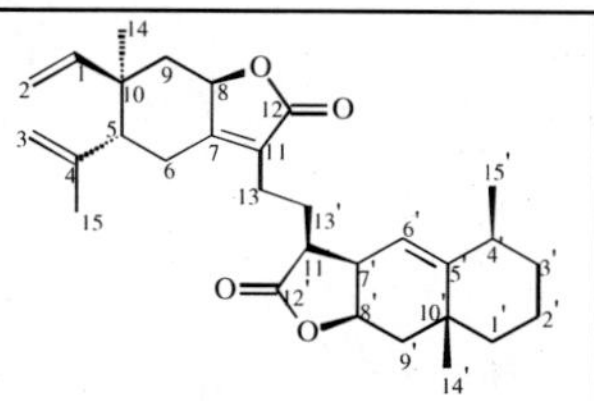	*I. macrophylla* (Su, B. N., et al., 2000)
2	Macrophyllidimer B $C_{30}H_{40}O_4$ Mr 464	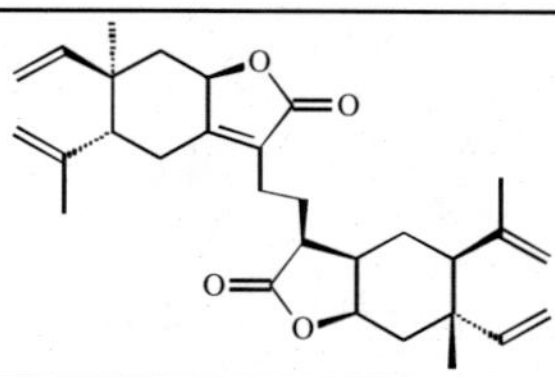	*I. macrophylla* (Su, B. N., et al., 2000)
3	Inulajaponicolide B $C_{32}H_{40}O_7$ Mr 536	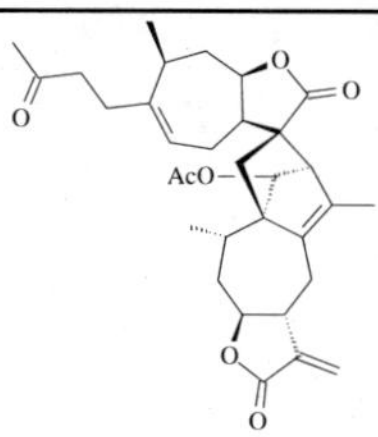	*I. japonica* (Jin, Q., et al., 2018)
4	Inulajaponicolide C $C_{30}H_{38}O_5$ Mr 478	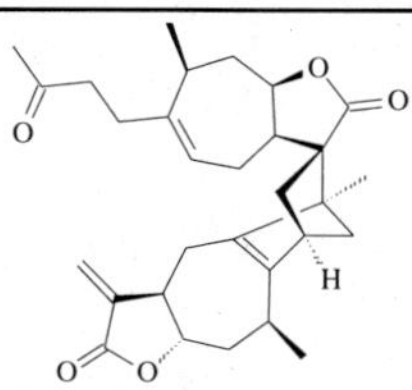	*I. japonica* (Jin, Q., et al., 2018)
5	Inulajaponicolide D C32H40O7 Mr 536	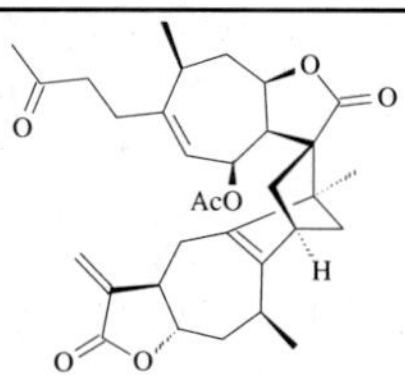	*I. japonica (Jin, Q., et al., 2018)*

Table 14. (Continued)

№	Sesquiterpene lactone	Structure	Source
6	Leariifolianoid E $C_{32}H_{42}O_8$ Mr 554 $(\alpha)_D^{20}$ - 7.2 (*c* 0.20, MeOH)	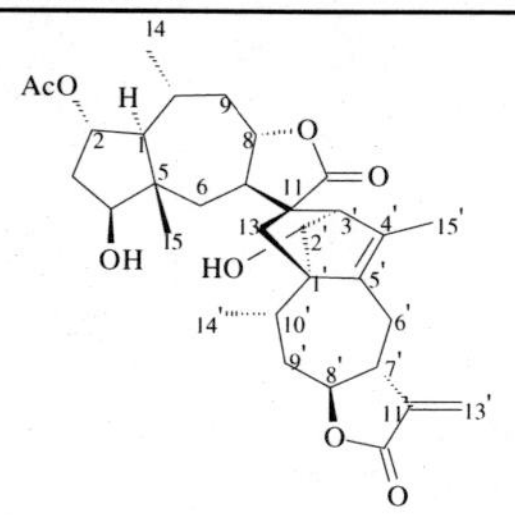	*I. lineariifolia* (Qin, J. J., et al., 2013)
7	Lineariifolianoid F $C_{34}H_{42}O_9$ Mr 594 $(\alpha)_D^{20}$ - 15.0 (*c* 0.10, MeOH)	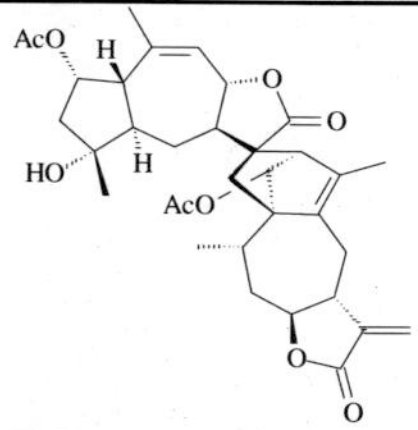	*I. lineariifolia* (Qin, J. J., et al., 2013)
8	Lineariifolianoid G $C_{32}H_{40}O_7$ Mr 536 $(\alpha)_D^{20}$ + 10.5 (*c* 0.20, MeOH)	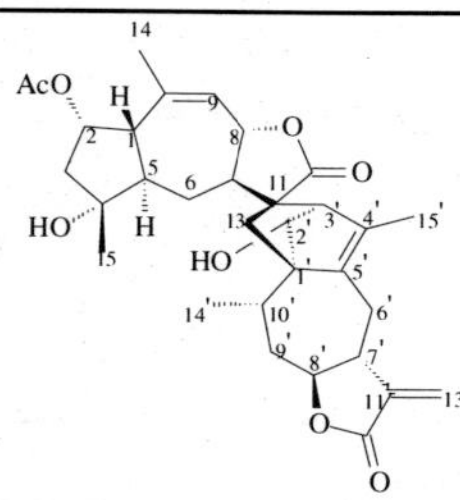	*I. lineariifolia* (Qin, J. J., et al., 2013)
9	Lineariifolianoids H $C_{32}H_{40}O_8$ Mr 552 $(\alpha)_D^{20}$ + 0.7 (*c* 0.20, MeOH)	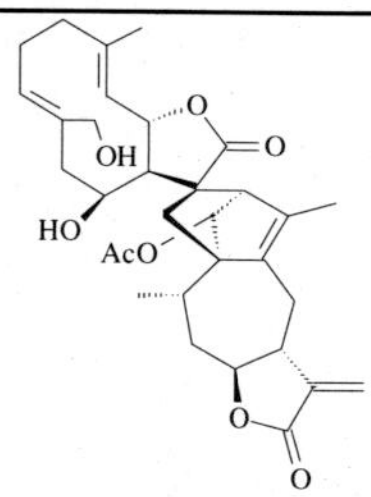	*I. lineariifolia* (Qin, J. J., et al., 2013)
10	lineariifolianoid I C32H42O8 Mr 577 (α)D25 -164.2 (c 0.048, CHCl3)	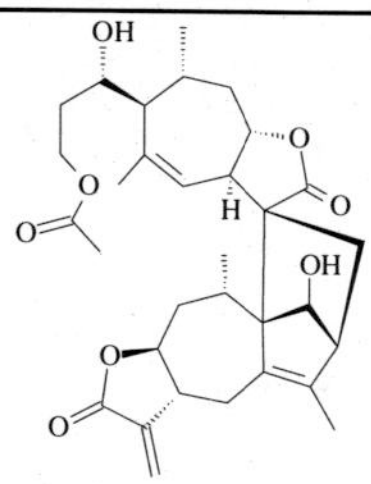	*I. lineariifolia (Chen L. p., et al., 2016)*

№	Sesquiterpene lactone	Structure	Source
11	lineariifolianoid J C34H42O10 Mr 633 (α)D25 +186.3 (c 0.062, CHCl3)	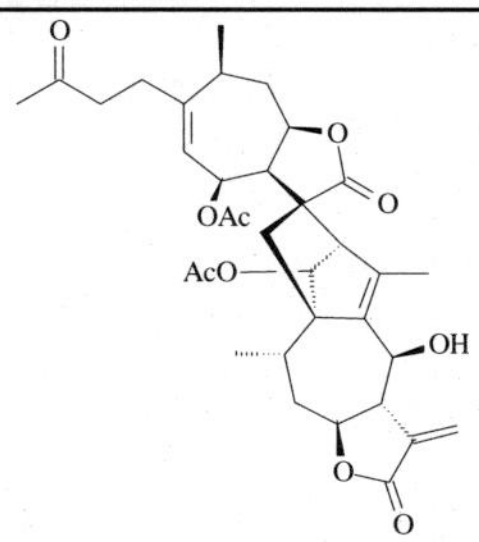	*I. lineariifolia (Chen L. p., et al., 2016)*
12	Lineariifolianoid K $C_{32}H_{40}O_8$ Mr 552	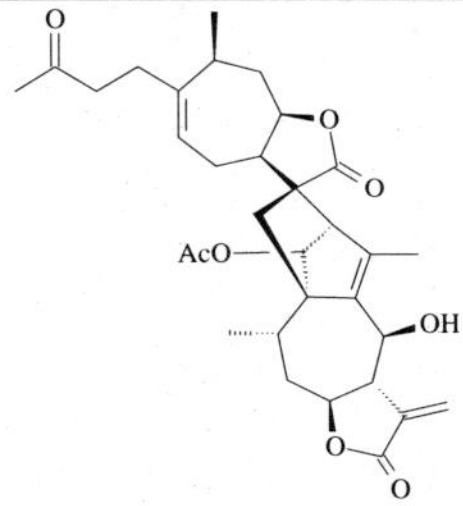	*I. lineariifolia* (Chen L. P., et al., 2016)
13	Lineariifolianoid L $C_{32}H_{40}O_8$ Mr 575 $(\alpha)_D^{25}$ +41.9 (*c* 0.062, $CHCl_3$)	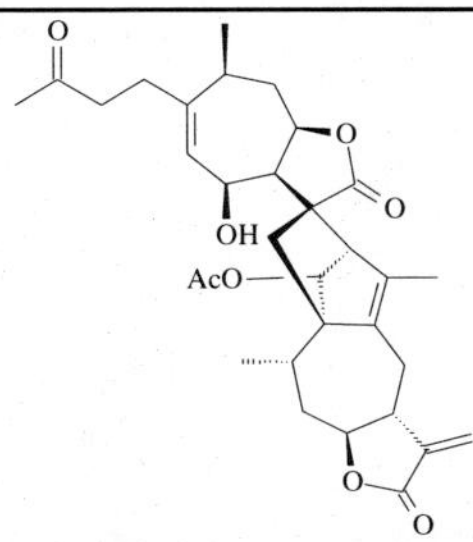	*I. lineariifolia* (Chen L. P., et al., 2016)
14	LINEARIIFOLIANOID M C34H42O9 MR 594 (α)D25 + 128.415	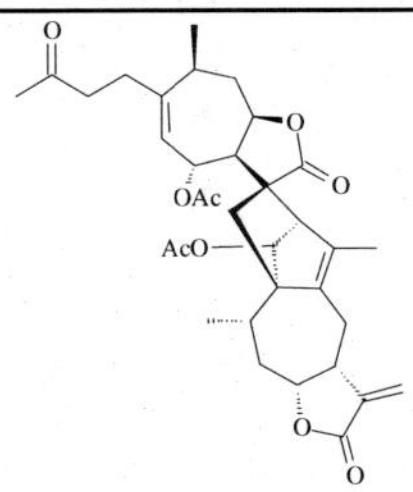	*I. lineariifolia (Su, R. x., et al., 2020)*
15	LINEARIIFOLIANOID N C34H44O9 MR 596 (α)D25 + 31.410	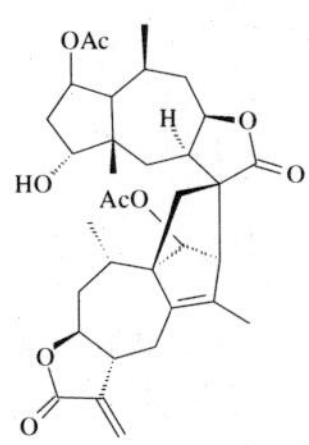	*I. lineariifolia (Su, R. x., et al., 2020)*

Table 14. (Continued)

№	Sesquiterpene lactone	Structure	Source
16	Leariifolianoid O $C_{32}H_{40}O_8$ Mr 552 $(\alpha)_D^{25}$ + 11.765	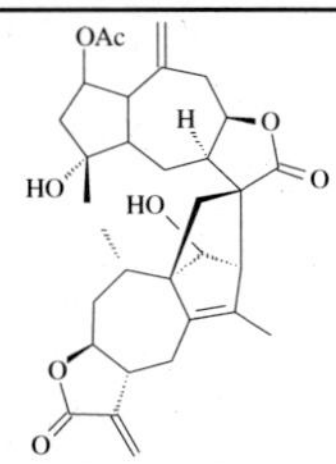	*I. leariifolia* (Su, R. X., et al., 2020)
Trimeric sesquiterpene lactone			
1	Inulajaponicolide A $C_{47}H_{56}O_9$ Mr 764	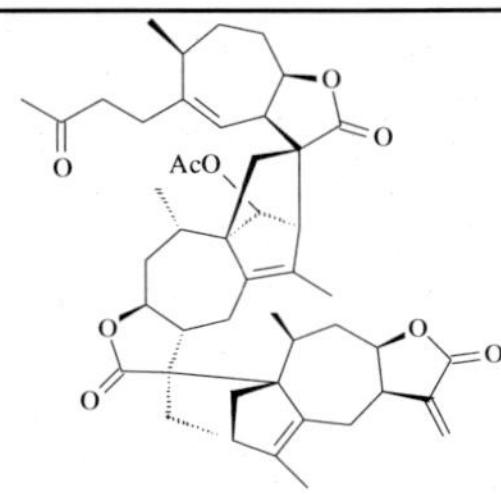	*I. japonica* (Jin, Q., et al., 2018)

According to the literature, 36 eudesmanolides, 20 germacranolides, 19 pseudoguaianolides, 16 guaianolides, 16 1,10-*seco*-eudesmanolides, 9 xanthanolides, 5 elemanolides, 2 4,5-*seco*-eudesmanolides, 2 eremophylanolides, 6 other types of lactones, 16 dimeric and 1 trimeric sesquiterpene lactones have been isolated from plants of the *Inula* L. genus over the past 5-6 years.

Biological Activity of Essential Oils and Extracts of the *Inula* L. Genus Plants

The components of essential oils are widely used in medicine, cosmetology and agriculture. Sesquiterpene lactones – is a large group of plant terpenoids with various biological activities, including: antitumor or cytotoxic, antiparasitic or anthelmintic, anti-inflammatory, hepatoprotective, hypolipidic, anti-atherosclerotic, antibacterial, antiviral, fungicidal, growth-stimulating, etc. (Matejić, J., Šarac, Z. and Ranđelović, V., 2010).

Biological Activity of Essential Oils and Non-Alcoholic Extracts of the *Inula* L. Genus Plants

The essential oil of *I. britannica* flowers, extracted using the supercritical fluid extraction method, is one of the potential and promising antifungal agents that can be used as a botanical fungicide to protect crops (Te, Z., Fei, G., Lin, Z., Tian-You, S., 2013).

Changes in electroencephalography values caused by inhalation of essential oil of *I. helenium* root can increase the state of alertness in a brain and therefore have a potential to treat psychophysiological disorders (Sowndhararajan, K., Choa, H., Yu, B., Song, J., Kim, S., 2016).

The essential oil of *Inula helenium* roots showed very strong antistaphylococcal activity against *Staphylococcus aureus*, due to the presence of sesquiterpene lactones of the eudesmanic type with an α,β-unsaturated lactone cycle (alantolactone, isoalantolactone and diprofillin) (Stojanović-Radić, Z., et al., 2012).

Inula viscosa root essential oil demonstrated strong antibacterial, anticholinesterase, anti-tyrosinase and anti-5-lipoxygenase potential (Aissa, I., et al., 2019).

Acetone extract of *Inula crithmoides* has significant antioxidant activity (IC_{50} = 13 mg/ml for DPPH) due to flavonoids. Essential oil is of lower antioxidant activity when compared to acetone extract; however, it shows the best antimicrobial action (Jallali, I., et al., 2014).

The bioluminescence test was used to study the antigenotoxic and antioxidant activity of biologically active substances of various extracts of *Inula britannica* on *E. coli* strains MG1655 (pColD-lux), MG1655 (pSoxS-lux), and MG1655 (pKatG-lux). The strongest antigenotoxic and antioxidant effects were found at a concentration of 500.0 μg/ml (Lovinskaya, A. V., et al., 2017).

Inula britannica extract, fermented by *Lactobacillus plantarum*, dose-dependently inhibited tyrosinase activity stimulated by α-melanocytes activating a hormone. These results indicate that fermented *I. britannica* extracts can be used as effective skin whitening agents (Park, E. H., et al., 2017).

Various *Inula cuspidata* extracts have shown significant antibacterial activity against Gram-positive bacteria with minimum inhibitory concentration (MIC) values ranging from 187.5 to 750 μg/ml and moderate to weak inhibition of Gram-negative bacteria *in vitro* with MIC values ranging from 750 to 3000 μg/ml. Research shows antioxidant and antibacterial activity of those extracts (Paliwal, S. K., Sati, Bh., Faujdar, S., Sharma, S., 2017).

The results of the study (Paliwal, S. K., Sati, Bh., Faujdar, S., Sharma, S., 2017) confirmed traditional application of *Inula cuspidata* as an analgesic and anti-inflammatory agent.

The antibacterial activity of *Inula graveolens* essential oil against *S. aureus* ATCC 6538P (CIP 53.156) has been reported elsewhere (Guinoiseau, E., et al., 2010).

Inula racemosa extract can be a potential therapeutic agent for providing several beneficial effects in hepatic ischemia-reperfusion injury (Wang, Z., et al., 2017).

Hexane extract of *Inula racemosa* (Hook) exhibited larvicidal and growth-inhibiting activity against larvae of the pest *Spodoptera litura* (Fabricius) (Kaur, M., et al., 2018).

Aqueous extract of *Inula viscosa* leaves exhibits antihypertensive effect (Kattouf, J., et al., 2009).

When studying extracts of various plants for wound healing activity, the best results were observed for an aqueous extract of *Inula viscosa* (Khalil, E. A., Afifia, F. U., Al-Hussaini, M., 2006).

In addition to pharmacological activity, some extracts have allelopathic properties. *Inula viscosa* leaf extracts have cytotoxic and genotoxic effects on cells of the root meristem of *Allium cepa* (Askin, T., Ozlem, C. Aslanturk, S., 2010).

Biological Activity of Alcoholic Extracts of the *Inula* L. Genus Plants

Ethanol extracts (50% and 70%) from the roots of the Romanian variety *I. helenium* L. showed significant antimicrobial activity against all tested microorganisms including *Bacillus subtilis, Bacillus cereus, Enterococcus faecalis, Escherichia coli, Staphylococcus aureus* and fungi *Candida albicans, C. parapsilosis, C. lipolytica*, except for pathogenic filamentous fungi *Aspergillus niger* (Digută, C., et al., 2014).

The alcoholic extract has an anti-inflammatory effect by inducing the p38 MAPK-dependent HO-1 signaling pathway (Park, E. J., et al., 2013).

Ethanol extract of *Inula helenium* elecampane rhizomes disrupts growth performance, dietary absorption, intestinal health and antioxidant status in broiler chickens (Abolfathi, M. E., et al., 2019).

The methanol extract of *Inula viscosa* showed an antihypertensive effect, mainly mediated by the endothelium-dependent vasodilating effect (Hakkou, Z., et al., 2017) while the ethanol extract of the leaves demostrated an antioxidant activity (Kheyar-Kraouche, N., et al., 2018).

Ethanol extract of *I. japonica* flowers *in vitro* modulates eicosanoid generation and degranulation through suppression of SCF-mediated signaling pathways. The later could be useful for preventing allergy in inflammatory diseases. Exhibitted Antiallergic activity can be in part due to the presence of britanin and tomentosin as major components as confirmed by HPLC analysis (Lu, Y., et al., 2012).

The methanol extract of *Inula graveolens* showed a significant anti-inflammatory effect at a dose of 400 mg/kg ($P < 0.01$) as compared with the standard drug diclofenac sodium (50 mg/kg) (Adnan, J., Al-Fartosy, M., 2013). The extract also demonstrated potential activity in inhibiting platelet aggregation. The maximum inhibition was observed at a dose of 400 μg/ml ($p < 0.01$) compared with heparin (20 μg/ml).

Biological Activity of Aqueous Extracts of the *Inula* L. Genus Plants

Inula britannica extract improves hyperglycemia and hepatic changes in streptozotocin-induced diabetic mice. Therefore, it can potentially be used as an antidiabetic supplement or drug (Zangeneh, M. Mahdi, et al., 2018).

Aqueous extract of *Inula crithmoides* has shown protective activity against oxidative stress caused by ochratoxin A, mutagenic effects, and also it has no toxicity (Abdel-Wahhaba, M. A., Abdel-Azim S. H., El-Nekeety A. A., 2008).

Aqueous (10, 20, 30 and 40 g/l) and organic (hexane, chloroform and methanol, at 3000 and 6000 ppm) extracts of *Inula crithmoides* (roots, stems, leaves and flowers) were evaluated for their allelopathological effect on radish, lettuce, pegan and thistle (Omezzinea, F., Ladhari, A., Rinez, A., Haouala, R., 2011).

Aqueous and organic extracts of *Inula ensifolia* were investigated *in vitro* to determine their antiproliferative activity against HeLa (adenocarcinoma of the cervical epithelium), A431 (epidermoid skin cancer) and MCF7 (adenocarcinoma of the mammary gland epithelial) cells using the MTT test (Rethyl, B., Csupor-Loffler, B., Zupko, I., Hajdu, Z., Mathe, I., Hohmann, J., Redei, T. Falkay, G., 2007).

Inula viscosa aqueous extract exhibitted anti-inflammatory potential both *in vivo* and *in vitro*. The effect demonstrated *in vitro* can be explained by the inhibition of lipopolysaccharide LPS-induced NO production due to expression inhibition of nitric oxide synthase-induced *i*NOS (Lounis, H., et al., 2018).

The aqueous extract of *Inula viscosa* has hypoglycemic activity in normal and diabetic rats, however, it does not have a hypolipidemic effect (Zeggwagh, N. A., Ouahidi, M. L., Lemhadri, A., Eddouks, M., 2006).

Inula racemosa aqueous extract has the strongest antioxidant and redox potential compared to ethanol and 50% aqueous ethanol extracts (Mohan, S., Gupta, D., 2017).

Biological Activity of Sesquiterpene Lactones of the *Inula* L. Genus Plants

Sesquiterpene lactones of plants of the *Inula* L. genus and their derivatives demonstrated a large spectrum of biological activities with dominant antitumor and anti-inflammatory properties.

Antitumor Activity of Sesquiterpene Lactones of the *Inula* L. Genus Plants

Sesquiterpene lactones of *Inula britannica* are biologically active substances with anti-tumor activity. They showed similar cytotoxic activity against drug-sensitive and drug-resistant cancer cell lines (Fischedick, J. T., et al., 2013).

For example, britanin has an antiproliferative effect on MCF-7 and MDA-MB-468 human breast cancer cells (Hamzeloo-Moghadam, M., et al., 2015).

1β-Hydroxyalantolactone from *Inula britannica* demonstrated anti-inflammatory and antitumor activity (Chen L., Zhang J. P., Liu X., Tang J. J., Xiang P., Ma X. M., 2017).

1-*O*-Acetylbritannilactone and 1,6-*O,O*-diacetylbritannilactone from *Inula britannica* induce cancer cell apoptosis and have anti-inflammatory effects (Ho, C. T.; et al., 2003; Pan, M. H., et al., 2007).

The anthracycline antibiotic daunorubicin was modified with the sesquiterpene lactones alantolactone, isoalantholactone and alloalantho-lactone from *Inula helenium*. The antiproliferative properties of the synthesized conjugates were extensively studied using tumor and normal cell lines. It was found that the conjugate daunorubicin with alantolactone showed stronger antiproliferative properties than daunorubicin without increasing toxicity to normal cells (Anikina, L. V., et al., 2018).

The bidirectional permeability of isoalantholactone and alantolactone from *Inula helenium* was investigated on monolayers of Caco-2 cells. The absorption permeability of isoalantholactone and alantolactone was good in the Caco-2 cell. The antiproliferative properties of conjugates of daunorubicin and doxorubicin (anthracycline antibiotics) with sesquiterpene lactones (alantolactone, isoalantholactone, and alloalantholactone) have been studied in tumor and regular cell lines. It was found that their epoxy derivatives exhibit higher activity against human tumor cell lines than the corresponding doxorubicin conjugates. Daunorubicin-conjugate with epoxyisoalantholactone turned out to be more cytotoxic than daunorubicin in a number of cell lines, including those resistant to daunorubicin, but not affecting regular human cells (Xu, R., Peng, Y., Wang, M., Li, X., 2019; Semakov, A. V., et al., 2018).

Alantolactone from *Inula helenium* exerts antitumor effects in human glioblastoma cells through signaling mediated by the NF-κB/COX-2 cascade and, therefore, inhibiting IKKβ kinase activity. Plant-derived small molecule inhibitor alantolactone, is a promising candidate for clinical application in the treatment of CNS tumors (Wang, X., et al., 2017).

Isoalantholactone from *Inula helenium* has antifungal and cytotoxic properties. The effect of isoalantholactone on cell viability, cell cycle and apoptosis has been screened against stomach of a human with SGC-7901 cell adenocarcinoma (Rasul, A., et al., 2013).

When compared to the used cytosine arabinoside (Ara-C), alantolactone showed higher cytotoxicity effects on leukemia, sparing normal hematopoietic cells. Alantolactone induced apoptosis in a dose-dependent manner by inhibiting NF-kB and its downstream target proteins. DMA-alantolactone, a water-soluble prodrug of alantolactone, could inhibit tumor growth *in vivo*. Alantolactone can be considered as a new therapy targeting LSC (leukemic stem cell) leukemia and eudesman sesquiterpene-type lactones offer a new scaffold for LSC drug discovery (Ding, Y., et al., 2016).

The data from paper (Lei, J. C., Yu, J. Q., Yin, Y., Liu, Y. W., Zou, G. L., 2012) suggest that alantolactone could potentially be elaborated as an agent to treat human liver cancer.

Isoalantolactone, septuplinolide, 4α,13-dihydroxy-5,7(11)-eudesma-dien-12,8-olide, macrophyllactone, 13-acetyloxy-5,7(11)-eudesmadien-12,8-

olide, 11α,13-dihydro-2α-hydroxyalantolactone, 11,13-dihydroivaline and 1-oxo-4-epi-alantolactone showed cytotoxic activity (Zhang, T., Gong, T., Yang, Y., Chen, R. Y., Yu, D. Q., 2012). Isoalantholactone might be applied as a chemotherapeutic agent for human breast cancer (MDA-MB-231 and MCF-7). Moreover, it found to be no toxic to healthy breast cells MCF-10A (Li, Z., Qin, B., Qi, X., Mao, J., Wu, D., 2016).

Isoalantolactone exhibits growth inhibitory activity against ESCC (esophageal squamous cell carcinoma) cells, which is explained by the activation of PDCD4 (programmed cell death protein 4) through the suppression of microRNA-21 (small non-coding RNA molecules 18-25 nucleotides in length) (Wen, S. W., et al., 2018).

Apoptosis of tumor cell lines was studied *in vitro* with the participation of amino derivatives of epoxyalantolactone exhibiting antitumor activity (Pukhova, S. A., et al., 2018).

Isocostunolide can effectively induce cytotoxicity in three cancer cell lines (A2058, HT-29 and HepG2) with IC_{50}s of 3.2, 5.0, and 2.0 mg/ml, respectively. This compound induces apoptosis through a mitochondria-dependent pathway in A2058 cells (Chen, C. N., et al., 2007).

Telekin strongly inhibits the proliferation of cancer cells. Moreover, it also demonstrates excellent antiproliferative activity in hepatocellular carcinoma cells and low cytotoxicity to normal hepatocyte cells (Zheng, B., et al., 2013).

5α-Epoxyalantolactone from *Inula helenium* induces apoptosis in KG1a cells. In addition, the compound had effected the levels of proteins associated with apoptosis (Bcl-2, Bax, cytochrome C, caspase 9 and caspase 3). This indicates that the antiproliferative effect of the compound on KG1a cells may be mediated through a mitochondria-dependent mechanism by apoptotic pathway (Ding, Y., et al., 2019).

Gaillardin from *Inula hookeri* might inhibit the proliferation of breast cancer cells by inducing the mitochondrial apoptotic pathway and, therefore, could be promising agent in chemoprophylaxis or chemotherapy (Fallahian, F., Aghaei, M., Abdolmohammadi, M. H., Hamzeloo-Moghadam, M., 2015).

Gaillardin from *Inula oculus-christi* induces apoptosis in breast cancer cell lines MCF-7 and MDA-MB-468, and the main mechanism of its antitumor action has been determined (Fallahian, F., Aghaei, M., Abdolmohammadi, M.H., Hamzeloo-Moghadam, M., 2015).

Induction of apoptosis by tomentosin and inuviscolide from *Inula viscosa* on human aggressive melanoma cell lines have high pharmacological value and imply that new agents for the treatment of melanoma can be developed from these lactones (Rozenblat, S., et al., 2008).

Bigelovin from *Inula helianthus-aquatica* is used in the treatment of cancer (leukemia U937) in Yunnan, China (Yue, G. G. L., et al., 2013). In addition, Bigelovin has anti-angiogenic and immunomodulatory activities, suggesting that Bigelovin may have multipurpose anti-cancer functions in animal models.

Bigelovin, diacetylovatifoline, 3β-hydroxy-11α,13-dihydro-alantolactone, 2α-acetoxy-4α,6α-dihydroxy-1β,5αH-guaia-9(10),11(13)-dien-12,8α-olide, britanlin G, inuchinenolide B, diacetylinuchinenolide B, showed *in vitro* activity against two human breast cancer cell lines (MCF-7 and MDA-MB-231) and one normal breast cell line (MCF-10A) (Qin, J. J., et al., 2013).

Lineariifolianoid E, lineariifolianoid F, lineariifolianoid G, lineariifolianoid H from *Inula lineariifolia* showed *in vitro* IC50 values of 1.56 mM and 2.75 mM against two human breast cancer cell lines MCF-7 and MDA-MB-231, respectively. In addition, lactone showed low toxicity to regular breast cells MCF-10A, indicating its selective cytotoxicity toward tumor cells. Further research also showed that lineariifolianoid E has a significant dose-dependent effect on cells (Qin, J. J., et al., 2013).

The authors (Tang, J. J., et al., 2018) carried out cytotoxic activity *in vitro* of 1β-hydroxyalantholactone, ivangoustine against six cell lines, including HeLa, PC-3, HEp-2, HepG2, CHO and HUVEC.

Japonicones A - D from *Inula japonica* have cytotoxic activity against 4 cell line tumors including A549, LOVO, CEM and MDA-MB-435 (Qin, J. J., et al., 2009).

Neojaponicone B and inulanolide E from *Inula japonica* showed potent cytotoxicity with an IC_{50} value of 2.2-5.9 μm (Xu, X. Y., et al., 2015).

Compounds (5*S*, 7*R*, 8*S*, 10*R*)-4-oxopseudoguai-1,11(13)-dien-12,8-olide, (1*S*, 2*R*, 5*S*, 7*R*, 8*S*, 10*R*)-2-methoxy-4-oxypseudoguai-11(13)-en-12,8-olide from *Inula japonica* was tested in a laboratory cytotoxic assay against human cancer cell lines A549, MCF-7 and MDA-MB-231 (Yu, Z. P., et al., 2019).

Sesquiterpene lactones from *Inula japonica* exhibited potent cytotoxicity against cancer cell lines such as HL-60, SMMC-7721, A-549, MCF-7, and SW-480 with IC_{50} values ranging from 1.57 to 22.58 mm. Several selected compounds also exhibited significant inhibitory activity against LPS-induced NO production in RAW264.7 macrophages with IC_{50} values ranging from 1.42 to 8.99 mm (Wu, X. D., et al., 2016).

Racemosalactones A - E exhibit antiproliferative activity with IC_{50} values ranging from 0.38 to 4.19 μg ml against human cell lung cancer A549, hepatocellular carcinoma HepG2 and human fibrosarcoma cells HT1080 (Ma, Y. Y., Zhao, D. G., Gao, K., 2013).

Ineupatolide B from *Inula cappa* has moderate inhibitory activity against human leukemia cells HL-60 (Wang, F. Y., et al., 2012).

Ineupatolide D, ineupatolide E from *Inula cappa* showed moderate inhibitory effect on cancer cell lines including A431, A549, BGC-823, HL-60, HT-29 and MCF-7 with IC_{50} values ranging from 2.1 to 36.3 μM (Wu, J. W., et al., 2017).

Anti-Inflammatory Activity of Sesquiterpene Lactones of the *Inula* L. Genus Plants

Isoalantolactone exhibits anti-inflammatory effects on lipopolysaccharide (LPS)-stimulated BV2 microglial cells through activation of the GSK-3β-Nrf2 signaling pathway (Wang, M., Wang, K., Gao, X., Zhao, K., Chen, H., Xu, M., 2018).

Alantolactone showed a significant effect on suppressing the release of IL-8, TNF-α and IL-1β, comparable to budesonide used as a positive control. *Inula helenium* is a valuable source of active compounds with anti-inflammatory properties (Gierlikowska, B., et al., 2020).

Alantolactone and isoalantolactone demonstrate the potential of TSL-IHL in the prevention and treatment of rheumatoid arthritis (Gaoa, S., et al., 2017).

The thiol analogs of alantolactone have therapeutic potential and could be further investigated for potential anti-inflammatory effects (Kumar, C., et al., 2019).

Bigelovin from *Inula britannica* L. var. *chinensis* inhibits the adhesion of inflammatory monocytes to endothelial cells and the expression of VCAM-1, ICAM-1, and E-selectin by blocking the degradation of IkBα and NF-kB activation (Nam, K. W., et al., 2009).

Hookerolides A - D from *Inula hookeri* have moderately anti-inflammatory and cytotoxic actions (Cheng, X. R., et al., 2013b).

Acetylbritannilactone is able to suppress abnormal proliferation of vascular smooth muscle cells accompanied by the induction of apoptosis *in vivo* and *in vitro*. Lactone can be further considered in pharmacology as a candidate for the prevention of restenosis after balloon angioplasty (Liu, B., et al., 2011).

Britanin may have potential to treat inflammatory diseases by inhibiting MAP kinases and NF-κB-mediated macrophage activation (Park, H. H., Kim, M. J., Li, Y., Park, Y. N., et al., 2013).

Pterocaullin B from *Inula pterocaula* shows potential anti-inflammatory activity at an IC_{50} value of 7.03 μm (Zhu, S. L., et al., 2019).

Compounds 4α,6α-dihydroxy-1β,5α,7αH-guaia-9(10),11(13)-dien-12,8α-olide, ergolide, 4α,5α-epoxy-10α,14H-1-*epi*-inuviscolide and granilin showed stronger inhibitory activity than the positive control with IC_{50} values of 0.13, 0.07, 0.11 and 0.11 mM, respectively. These studies have also contributed to a better understanding of the structure-activity relationship for sesquiterpene lactones (Cheng, X., Zeng, Q., Ren, J., Qin, J. J., et al., 2011).

JEUD-38 from *Inula japonica* is a promising candidate for the prevention and treatment of inflammatory diseases (Wang, X., Tang, Sh. An, Wang, R., Qiu, Y., et al., 2015).

Sesquiterpene lactones from *Inula wissmanniana*: (5*R*,7*R*,10*S*)-4,5-epoxy-4,5-*seco*-eudesma-1,3-dien-12,5-olide, eudesma-3,11(13)-dien-12,5β-olide, (2*R*, 4*E*, 6*R*, 7*S*, 9*S*, 10*R*)-2-acetoxy-9-angeloyloxygermacra-4,11 (13)-dien-12,6-olide, 4*E*-9β-angeloyloxy-2α-hydroxy-7α,10αH-germacra-4,11(13)-dien-12,6α-olide, (4*E*, 6*R*, 7*S*, 9*S*, 10*R*)-9-angeloyloxy-germacra-4,11(13)-dien-12,6-olide, 4*E*-9β-methacryl-oxy-7α,10αH-germacra-4,11(13)-dien-12,6α-olide, 4*E*-9β-hydroxy-7α,10αH-germacra-4,11(13)-dien-12,6α-olide, 4β,5α-dihydroxy-1-oxoeudesma-2,11(13)-dien-12,6β-olide, 3-oxoeudesma-1,4,11(13)-triene-12,6β-olide possess activity to inhibit lipopolysaccharide-induced nitric oxide production in macrophages RAW264.7 and cytotoxicity in tumor cells HepG2, PC-3 and MGC-803 (Cheng, X. R., et al., 2013a).

Two new germacranolides (4*R*,5*R*,6*S*,7*S*,9*S*,10*R*)-9-angeloyloxy-4,5-epoxygermacra-11(13)-en-12,6-olide and 4*E*-9β-angeloyloxy-7α,10α, 11αH-germacra-4,11(13)-dien-12,6α-olide from *Inula wissmanniana* were evaluated for their inhibition against LPS-induced nitric oxide (NO) production in macrophages RAW264.7. (4*R*, 5*R*, 6*S*, 7*S*, 9*S*, 10*R*)-9-Angeloyloxy-4,5-epoxygermacra-11(13)-en-12,6-olide showed the strongest activity with an IC_{50} value of 1.04 μM (Cheng, X. R., Wang, C. H., Wei, P. L., Zhang, X. F., et al., 2014).

Leariifolianoides M - O showed a significant inhibitory effect on LPS-induced production in RAW 264.7 macrophages with IC_{50}s of 1.421, 1.087 and 1.243 μm, respectively (Su, R. X., et al., 2020).

Tomentosin, inuchinenolide A, 6β-hydroxytomentosin, 8β-propionyloxy-inusoniolide showed inhibition of LPS-induced nitric oxide production in macrophages RAW264.7 (Nie, L. Y., et al., 2010).

1,6α-Dihydroxy-4α*H*-1,10-*seco*-eudesma-5(10),11(13)-dien-12,8β-olide, 6α-isobutyryloxy-1-hydroxy-4α*H*-1,10-*seco*-eudesma-5(10),11(13)-dien-12,8β-olide, 6α-isovaleryloxy-1-hydroxy-4α*H*-1,10-*seco*-eudesma-5(10),11 (13)-dien-12,8β-olide has an inhibitory effect on nitric oxide production in lipopolysaccharide (LPS)-stimulated RAW264.7 cells (Chen, X., et al., 2015).

Compounds hupehenolide B, hupehenolide E, neohupehenolide A, neohupehenolide B, hupehenolide M, britanin, gaillardin, 2α-acetoxy-inuviscolide, 5α,6α-epoxy-2α-acetoxy-4α-hydroxy-1β,7α(H)-guaia-1(13)-en-12,8α-olide, loliolide and dehydrovomifoliol were evaluated for their inhibitory effect on LPS-induced NO production in RAW264.7 macrophages. Compounds hupehenolide E, britanin, 2α-acetoxy-inuviscolide and 5α,6α-epoxy-2α-acetoxy-4α-hydroxy-1β,7α-guai-1(13)-en-12,8α-olide demonstrated significant inhibitory activity with values IC_{50} in the range of 0.6-6.6 μM (Ren, J., et al., 2013).

Other Types of Activity of Sesquiterpene Lactones of *Inula* L. Genus Plants

Inuviscolide has an inhibitory effect on elastase, cyclooxygenase 1 and secretory phospholipase A2. In addition, it reduces the infiltration of skin leukocytes in a mouse model of dermatitis induced by repeated application of 13-acetate-12-*O*-tetradecanoylphorbol (Manez, S., et al., 2007).

Japonicone V from *Inula japonica* inhibits hepatitis E virus replication by targeting virus-associated autophage (Zhao, Y., et al., 2020).

Inuloxines A-D from *Inula viscosa* have phytotoxic activity (Andolfi, A., et al., 2013).

Tomentosin and britanin from *Inula japonica* exhibit anti-allergic activity (Lu, Y., et al., 2012).

1β-Hydroxyalantolactone из метанольного экстракта *Inula japonica* может быть потенциальным агентом для лечения идиопатического легочного броза (Li, X., et al., 2020).

The lactones dehydrocostuslactone and costunolide and the crude extract of *Inula racemosa* were tested for antimicrobial properties and the pure compounds were evaluated for their ability to inhibit melanin synthesis (Venkateswara, Rao G., et al., 2013).

Isoalantolactone showed absolute toxicity at 500 μg/ml against 3 soil phytopathogenic fungi (*Gaeumannomyces graminis* var. *tritici, Rhizoctonia cerealis* and *Phytophthora capsici*) with MICs defined as 100, 100 and 300 μg/ml, respectively. In addition, isoalantholactone exhibits weaker antibacterial activity against *Bacillus subtilis, Escherichia coli, Pseudomonas fluorecense, Sarcina lentus* and *Staphylococcus aureus* with MICs of 125, 425, 150, 150 and 100 μg/ml, respectively. The study showed strong inhibition of phytopathogenic fungi by isoalantholactone (Changhong, L., Mishra, A. K., Bing, H., Renxiang, T., 2001).

Isoalantolactone has shown estrogenic (Kalachaveedu, M., et al., 2018), repellent, insecticidal and phytotoxic activity (Liu, C. H., Mishra, A. K., Tan, R. X., 2006).

Sesquiterpene lactones (3a*S*,8a*R*)-6-methyl-3-methyliden-7-(3-oxobutyl) -3a,4,5,8a-tetrahydro-2H-cyclohepta(b)furan-2,8(3H)dione and 9β-hydroxy-parthenolide demonstrated the strongest activity against (MIC $3{\bullet}10^{-2}$ - $5{\bullet}10^{-1}$ μg/ml), while (3a*S*,8a*R*)-6-methyl-3-methyliden-7-(3-oxobutyl)-3a,4,5,8a-tetrahydro-2H-cyclohepta(b)fu-ran-2,8(3H)dione and tomentosin turned out to be the most active against the group of analyzed fungi (MIC $2.5{\bullet}10^{-2}$ - $12{\bullet}10^{-2}$ μg/ml). The antiproliferative effect of lactone was also evaluated *in vitro* against the cell line of human bronchopulmonary cancer NSCLC-N6 and the epidermoid line of lung cancer A549 (Michalakea, E., et al., 2019).

Conclusion

Thus, it has been shown that terpenoids are a promising class of natural compounds for their extensive application in medicine, perfumery and agriculture. In depth scientific research provides strong evidence on plants of the genus *Inula* L. serving as a rich source for biologically active terpenoids.

References

Abdel-Wahhaba, M. A., Abdel-Azim S. H., El-Nekeety A. A. (2008). "*Inula crithmoides* extract protects against ochratoxin A-induced oxidative stress, clastogenic and mutagenic alteractions in male rats." *Toxicon*, 52: 566-573.

Abolfathi, M. E., Tabeidian, S. A., Foroozandeh Shahraki, A. D., Tabatabaei, S. N., Habibian, M. (2019). "Effects of ethanol extract of elecampane (*Inula *helenium* L.) rhizome on growth performance, diet digestibility, gut health, and antioxidant status in broiler chickens." *Livestock Science*, 223: 68-75.

Adekenov S. M., Abdykalykov M. A., Turdybekov K. M., Struchkov Yu. T., Pushin A. N. (1990). "Pulchellin C and inuchinenolide C from *Inula caspica*." *Chemistry of Natural Compounds*, 26: 635-642. https://doi.org/10.1007/BF00630071.

Adekenov, S. M., Kadirberlina, G. M., Turdybekov, K. M., Struchkov Yu. T. (1991b). "Terpenoids of *Crepis tectorum*: Molecular and crystal structure of the sesquiterpene lactone 8-epideacylcynaropicrin." *Chemistry of Natural Compounds*, 27: 562-565. https://doi.org/10.1007/BF00630355.

Adekenov, S. M., Turdybekov, K. M., Kadirberlina, G. M., Lindeman S. V., Struchkov Yu. T. (1991). "Molecular and crystal structure of jurineolide - A germacrane lactone from *Jurinea multiflora*." *Chemistry of Natural Compounds*, 27: 426-429. https://doi.org/10.1007/BF00636561.

Adekenov, S.M., Turdybekov, K.M., Aituganoav, K.A., Lindeman S. V., Struchkov Yu. T., Shaltakov S. N. (1993). "1β,10α-Dihydroxyarglabin - A new sesquiterpene lactone from *Artemisia glabella*." *Chem. Nat. Compd.*, 29: 735-739. https://doi.org/10.1007/BF00629642.

Adnan, J., Al-Fartosy, M. (2013). "Some pharmacological studies on the methanolic extract of *Inula graveolens* L." *J. Biomedical Science and Engineering*, 6: 1040-1049. http://dx.doi.org/10.4236/jbise.2013.611130.

Afifi, F. U., Abu-Dahab, R., Kasabri, V. and Abaza, I. M. (2015). "GC-MS composition and antiproliferative activity of *Inula graveolens* (L.) Desf. essential oil." *Arabian Journal of Medicinal & Aromatic Plants*. 1 (1): 57-66.

Ahmed, A., Mohamed, A. H., Tzakou, O., Petropoulou, A., Hassan, M. E., El-Maghraby, M. A., Zeller, K. P. (2003). "Terpenes from *Inula verbascifolia*." *Phytochemistry*. 62 (8): 1191-1194.

Aissa, I., Znati, M., Zardi-Bergaoui, A., Flamini, G., Ascrizzi, R., Ben Janne, H. (2019). "GC and GC-MS integrated analyses and *in vitro* antibacterial, anticholinesterase, antityrosinase, and anti-5-lipoxygenase potential of *Inula viscosa* root fractionated essential oil." *South African Journal of Botany*, 125 (September): 386-392.

Alalan, L., Al-Shammaa, I., Al-nouri, A. S. (2016). "Isolation and identification of six terpenes extracted from leaves of *Inula viscosa* (L.) grown in Syria" *Journal of Chemical and Pharmaceutical Research*. 8 (1): 27-31.

Aljancic, I. S., Vajs, V. E., Tesevic, V. V., Milosavljevic, Some, S. M. (2008). "Some Wild-Growing Plant Species from Serbia and Montenegro as the Potential Sources of Drugs. "*Current Pharmaceutical Design*, 14: 3089-3105.

Al-Snafi, Ali Esmail (2018). "Chemical constituents and pharmacological effect of *Inula graveolens* (syn: *Dittrichia graveolens*)." *IAJPS* 5 (4): 2183-2190.

Andolfi, A., Zermane, N., Cimmino, A., Avolio, F., Boari, A., Vurro, M., Evidente, A. (2013). "Inuloxins A–D, phytotoxic bi-and tricyclic sesquiterpene lactones produced by *Inula viscosa*: Potential for broomrapes and field dodder management." *Phytochemistry*, 86: 112–120.

Anikina, L. V., Semakov, A. V., Afanas'eva, S. V., Pukhov, S. A., Klochkov, S. G. (2018). "Synthesis and antiproliferative activity of daunorubicin conjugates with sesquiterpene lactones." *Pharmaceutical Chemistry Journal*, 52 (4): 308-311 (Russian Original Vol. 52, No. 4, April, 2018) https://doi.org/10.1007/s11094-018-1812-7.

Askin, T., Ozlem, C. Aslanturk, S. (2010). "Evaluation of Cytotoxicity and Genotoxicity of *Inula viscosa* Leaf Extracts with *Allium* Test." *Journal of Biomedicine and Biotechnology*, Volume 2010, Article ID 189252, 8 pages http://dx.doi.org/10.1155/2010/189252.

Beghidja, N., Ikhlef, F., Benayache, S., Benayache, F., Bouheroum, M. J., Chalchat, C. (2014). "Composition of the essential oil of *Inula graveolens* Algerian origin species." *Journal Natural Product Plant Resources*. 4 (1): 1-3.

Kattouf, J., Belmoukhtar, M., Harnafi, H., Mekhfi, H., Ziyyat, A., Aziz, M., Bnouham, M., Legssyer, A. (2009). "Antihypertensive effect of an aqueous extract of *Inula viscosa* leaves." *Phytotherapie*, 7: 309-312 (Kattouf, J., Belmoukhtar, M., Harnafi, H., Mekhfi, H., Ziyyat, A., Aziz, M., Bnouham, M., Legssyer, A. (2009). "Effect antihypertenseur des feuilles d'*Inula viscosa*." *Phytothérapie*, 7: 309-312).

Bo, Fu, Bao-Ning, Su, Takaishi, Y., Honda, G., Ito, M., Takeda, Y., Kodzhimatov, O. K., Ashurmetov, O. (2001). "A bis-sesquiterpene and sesquiterpenolides from *Inula macrophylla*." *Phytochemistry*, 58: 1121-1128.

Boudouda, H. B., Kabouche, A., Aburjai, T., Kabouche, Z. (2013). "GC-MS Analysis of *Inula graveolens* (L.) Desf. from Algeria." *Journal of essential oil-bearing plants* (JEOP). 16 (5): 651-654. https://doi.org/10.1080/0972060X.2013.854486.

Changhong, L., Mishra, A. K., Bing, H., Renxiang, T. (2001). "Antimicrobial activities of isoalantolactone, a major sesquiterpene lactone of *Inula racemosa*." *Chinese Science Bulletin*, 46 (6): 498-501. https://doi.org/10.1007/BF03187267.

Chen L., Zhang J. P., Liu X., Tang J. J., Xiang P., Ma X. M. (2017). "Semisynthesis, an Anti-Inflammatory Effect of Derivatives of 1β-Hydroxyalantolactone from *Inula britannica*." *Semisynthesis, an Anti-Inflammatory Effect of Derivatives*, 22 (11): 1835.

Chen L. P., Wu G. Zh., Dong H. Yu., Yang N., Zhang W. D., Li H. L. (2016). "Lineariifolianoids I-L, four rare sesquiterpene lactone dimers inhibiting NO production from *Inula leaeariifolia.*" *RSC Advances*, **6:** 103296-103298. https://doi.org/10.1039/C6RA23092J.

Chen, C. N., Huang, H. H., Wu, C. L., Lin, C. P. C., Hsu, J. T. A., Hsieh, H. -P., Chuang, S. E., Lai, G. M. (2007). "Isocostunolide, a sesquiterpene lactone, induces mitochondrial membrane depolarization and caspase-dependent apoptosis in human melanoma cells." *Cancer Letters*, 246: 237-252. http://dx.doi.org/10.1016/j.canlet.2006.03.004.

Chen, X., Ji, N., Qin, N., Tang, S. A., Wang, R., Qiu, Y., Duan, H., Kong, D. and Jin, M. (2017). "6-*O,O*-Diacetylbritannilactone Inhibits Eotaxin-1 and ALOX15 Expression Through Inactivation of STAT6 in A549 Cells." *Inflammation*, 40 (6): 1967-1974.

Chen, X., Tang, S. A., Lee, E., Qiu, Y., Wang, R., Duan, H. Q., Dan, S., Jin, M., Kong, D. (2015). "IVSE, isolated from *Inula japonica*, suppresses LPS-induced NO production via NF-κB and MAPK inactivation in RAW264.7 cells." *Life Sciences*, 1 (124): 8-15.

Cheng, X., Zeng, Q., Ren, J., Qin, J. J., Zhang, S., Shen, Y., Zhu, J., Zhang, F., Chang, R., Zhu, Y., Zhang, W., Jin, H. (2011). "Sesquiterpene lactones from *Inula falconeri*, a plant endemic to the Himalayas, as potential anti-inflammatory agents." *European Journal of Medicinal Chemistry*, 46: 5408-5415.

Cheng, X. R., Ren, J., Wang, C. H., Guan, B., Qin, J. J., Yan, S. K., Jin, H. Z., Zhang, W. D. (2013b). "Hookerolides A–D, the first naturally occurring C17-pseudoguaianolides from *Inula hookeri.*" *Tetrahedron Letters,* 54: 1943-1946.

Cheng, X. R., Wang, C. H., Wei, P. L., Zhang, X. F., Zeng, Q., Yan, S. K., Jin, H. Z., Zhang, W. D. (2014). "New sesquiterpenic acids from *Inula wissmanniana.*" *Fitoterapia*, 95: 139-146.

Cheng, X. R., Wang, C. H., Wie, P. L., Zhang, X. F., Zeng, Q., Yan, S. K., Jin, H. Z., Zhang, W. D. (2014). "New sesquiterpenic acids from *Inula wissmanniana.*" *Fitoterapia*, 95: 139-146.

Cheng, X.-R., Zhang, S. D., Wang, C. H., Ren, J., Qin, J. J., Tang, X., Shen, Y. H., Yan, S. K., Jin, H. Z., Zhang, W. D. (2013a). "Bioactive eudesmane and germacrane derivatives from *Inula wissmanniana* Hand. Mazz." *Phytochemistry*, 96: 214-222.

Chun, J. K., Seo, D. W., Ahn, S. H., Park, J. H., You, J. S., Lee, C. H., Lee, J. C., Kim, Y. K., Han, J. W. (2007). "Suppression of the NF-kappaB signalling pathway by ergolide, sesquiterpene lactone, in HeLa cells." *J. Pharmacy Pharmacology*. 59: 561-566.

Digută, C., Cornea, C. P., Ioniţă, L, Brînduşe, E., Farcaş, N., Bobit, D., Matei, F. (2014). "Studies on antimicrobial activity of *Inula helenium* L. Romanian cultivar." *Romanian Biotechnological Letters*, 19 (5): 9699-9704.

Ding, Y., Gao, H., Zhang, Yu., Li, Y., Vasdev, N., Gao, Y., Chen, Y., Zhang, Q. (2016). "Alantolactone selectively ablates acute myeloid leukemia stem and progenitor cells." *Journal of Hematology & Oncology*, 9 (1): 93. https://doi.org/10.1186/s13045-016-0327-5.

Ding, Y., Pan, W., Xu, J., Wang, T., Chen, T., Liu, Zh., Xie, Ch., Zhang, Q. (2019). "Sesquiterpenoids from the roots of *Inula helenium* inhibit acute myelogenous leukemia progenitor cells." *Bioorganic Chemistry*, 86: 363-367. https://doi.org/10.1016/j.bioorg.2019.01.055.

Dusmatova, D. E., Terent'eva, E. O., Bobakulov, Kh. M., Mukhamatkhanova, R. F., Tsai, T. A., Tashkhodzhaev, B., Sham'yanov, I. D., Azimova, Sh. S., Abdullaev, N. D. (2019). "Nonpolar Constituents of *Inula grandis* Roots. Cytotoxic Activity of Igalan." *Chemistry of Natural Compounds*, 55 (3): 571-574. https://doi.org/10.1007/s10600-019-02747-y.

Dusmatova, D. E., Turgunov, K. K., Bobakulov, Kh. M., Mukhamatkhanova, R. F., Sham'yanov, I. D., Tashkhodzhaev, B. and Abdullaev, N. D. (2020). "New sesquiterpene lactone from *Inula britannica.*" *Chemistry of Natural Compounds*, 56 (5): 852-854.

Dusmatova, D. E., Uzbekov, V. V., Mukhamatkhanova, R. F., Beshko, N. Yu., Dzhuraeva, L. T. (2018). "Quantitative determination of sesquiterpene lactones britannin and inuchinenolide C in flowers of *Inula britannica* L. by HPLC". *Universum: Chemistry and Biology*, 7(49). URL: http://7universum.com/ru/nature/ archive/item/6089. (Дусматова, Д. Э., Узбеков, В. В., Мухаматханова, Р. Ф., Бешко, Н. Ю., Джураева, Л. Т. (2018). "Количественное определение сесквитерпеновых лактонов британина и инучиненолида С в цветках девясила британского *Inula britannica* L. методом ВЭЖХ." *Universum: Химия и биология*, 7(49). URL: http://7universum.com/ru/nature/ archive/item/6089).

Eliel, E. L., Wilen, S. H. *Stereochemistry of Organic Compounds* (New York-London-Sydney-Toronto: J. Wiley, 1994), 1296.

Fallahian, F., Aghaei, M., Abdolmohammadi, M. H., Hamzeloo-Moghadam, M. (2015). "Molecular mechanism of apoptosis induction by Gaillardin, a sesquiterpene lactone, in breast cancer cell lines Gaillardin-induced apoptosis in breast cancer cell lines." *Cell Biol Toxicol*, 31: 295-305. https://doi.org/10.1007/s10565-016-9312-6.

Fazylova, A. S., Turdybekov, K. M., Kadirberlina, G. M., Rakhimova B. B., Adekenov S. M. (2000). "Molecular structure of crepidioside A and isolipidiol from *Crepis multicaulis*." *Chemistry of Natural Compounds*, 36: 177-180. https://doi.org/10.1007/BF02236425.

Fischedick, J. T., Pesic, M., Podolski-Renic, A., Bankovic, J., de Vos, R.C.H., Peric, M., Todorovic, S., Tanic, N. (2013). "Cytotoxic activity of sesquiterpene lactones from *Inula britannica* on human cancer cell lines."

Phytochemistry Letters, 6: 246–252. https://doi.org/10.1016/j.phytol.2013.02.006.

Fischer, N. H., Olivier, E. J., Fischer, H. D. "*The biogenesis and chemistry of sesquiterpene lactones*". In *Progress in the Chemistry of Organic Natural Products,* edited by Herz, W., Grisebach, H., Kirby, G. W. 47-390. New York: Springer-Verlag, 1979.

Gaoa, S., Wanga, Q., Tian, X. H., Li, H. L., Shen, Y. H., Xua, X. K., Wu, G. Z., Hu, Z. L., Zhang, W. D. (2017). "Total sesquiterpene lactones prepared from *Inula helenium* L. has potentials in prevention and therapy of rheumatoid arthritis." *Journal of Ethnopharmacology*, 196: 39-46.

Ghedira, K., Goetz, P., Le, R. Jeune, (2011). "*Inula helenium* L. (Asteraceae): *Aunée*." *Phytotherapie*, 9: 176-179.

Giamperi, L., Bucchini, A., Fraternale, D., Genovese, S., Curini, M., Ricci, D. (2010). "Composition and antioxidant activity of *Inula crithmoides* essential oil grown in central Italy (Marche region)." *Natural Product Communnications*, 5 (2): 315-318.

Gierlikowska, B., Gierlikowski, W., Bekier, K., Skalicka-Woźniak, K., Czerwińska, M. E., Kiss, A. K. (2020). "*Inula helenium* and *Grindelia squarrosa* as a source of compounds with anti-inflammatory activity in human neutrophils and cultured human respiratory epithelium." *Journal of Ethnopharmacology*, 249: 112311.

Goodwin, T. W. "*The Biogenesis of Terpenes and Steroids*." In *Rodd's Chemistry of Carbon Compounds* (Second Edition) In *A Modern Comprehensive Treatise*, edited by Elsevier BV, Chapter 19 (II): 54-137. Place: Publisher, 1964. https://doi.org/10.1016/B978-044453345-6.50248-0.

Guinoiseau, E., Luciani, A., Rossi, P. G., Quilichini, Y., Ternengo, S., Bradesi, P., Berti, L. (2010). "Cellular effects induced by *Inula graveolens* and *Santolina corsica* essential oils on *Staphylococcus aureus*." *Eur. J. Clin. Microbiol. Infect. Dis*., 29: 873–879. https://doi.org/10.1007/s10096-010-0943-x.

Hakkou, Z., Maciuk, A., Leblais, V., Bouanani, N. E., Mekhfi, H., Bnouham, M., Aziz, M., Ziyyat, A., Raufd, A., Hadda, T. B., Shaheen, U., Patel, S., Fischmeister, R., Legssyer, A. (2017). "Antihypertensive and vasodilator effects of methanolic extract of *Inula viscosa*: Biological evaluation and POM analysis of cynarin, chlorogenic acid as potential hypertensive." *Biomedicine & Pharmacotherapy*, 93: 62-69.

Hamzeloo-Moghadam, M., Aghaei, M., Fallahian, F., Jafari, S. M., Dolati, M., Abdolmohammadi M. H., Hajiahmadi S., Esmaeili S. (2015). "Britannin, a sesquiterpene lactone, inhibits proliferation and induces apoptosis through the mitochondrial signaling pathway in human breast cancer cells." *Tumor Biol.*, 36: 1191-1198. https://doi.org/10.1007/s13277-014-2744-9.

Hana, M., Wena, J. K., Zhenga, B., Zhang, D. Q. (2004). "Acetylbritannilatone suppresses NO and PGE2 synthesis in RAW 264.7 macrophages through the inhibition of iNOS and COX-2 gene expression." *Life Sciences*, 75: 675–684.

Harzallah-Skhiri, F., Cheraif, I., Jannet, H. B., Hammami, M. (2005). "Chemical composition of essential Oils from Leaves-stems, flowers and Roots of *Inula graveolens* from Tunisia." *Pakistan Journal of Biological sciences*, 8 (2): 249-254.

Hernandez-Ochoa, L., Vilarem, G., Mouloungui, Z., Medina-Gonzalez, Y. (2012). "Comparing the effect of ethyl heptanoate as co-solvent during extracting of essential oil of *Inula helenium* in a hydradistillation process." *Chemistry of Natural Compounds*, 47 (6): 996.

Ho, C. T.; Rafi, M.; Bai, N.; Dipaola, R. S.; Ghai, G.; Rosen, R. T. (2003). "Inducing cell apoptosis and treating cancer using 1-*O*-acetylbritannilactone or 1,6-*O*,*O*-diacetylbritannilactone." *US Pat*., (6): 623-627.

http://www.chimfak.sfedu.ru/

http://www.theplantlist.org/

https://studref.com/532205/matematika_himiya_fizik/tsikly_bolshego_razmera

Hua, Y., Qin, J., Zhang, F., Cheng, X. R., Jin, H., Zhang, W. (2012). "Sesquiterpene lactones from *Inula helianthus-aquatica." China J. Chin. Mater. Med*., 37 (11): 1586–1589.

Ivasenko, S. A., Edil'baeva, T. T., Kulyyasov, A. T., Atazhanova G. A., Drab A. I., Turdybekov K. M., Raldugin V. A., Adekenov S. M. (2006). "Structure and biological activity of α-santonin chloro-derivatives." *Chemistry of Natural Compounds*, 42: 36-40. https://doi.org/10.1007/s10600-006-0031-8.

Izbosarov, M. B., Abduazimov, B. K., Yusupova, I. M., Tashkhodzhaev, Vdovin A. D., Abdullaev N. D. (1998). "Germacranolide from *Tanacetopsis mucronata*." *Chemistry of Natural Compounds*, 34: 456-461. https://doi.org/10.1007/BF02329594.

Jallali, I., Zaouali, Y., Missaoui, I., Smeoui, A., Abdelly, C., Ksour, R. (2014). "Variability of antioxidant and antibacterial effects of essential oils and acetonic extracts of two edible halophytes: *Crithmum maritimum* L. and *Inula crithmoides* L." *Food Chemistry*, 145: 1031-1038.

Jiang, H. L., Chen, J., Jin, X. J., Yang, J. L., Li, Y., Yao, X. J., Wu, Q. X. (2011). "Sesquiterpenoids, alantolactone analogues, and seco-guaiene from the roots of *Inula helenium*." *Tetrahedron*, 67: 9193-9198.

Jin, Q., Lee, J. W., Jang, H., Lee, H. L., Kim, J. G., Wu, W., Lee, D., Kim, E. H., Kim, Y., Hong, J. T., Lee, M. K., Hwang, B. Y. (2018). "Dimeric- and trimeric sesquiterpenes from the flower of *Inula japonica." Phytochemistry*, 155: 107-113.

Kagarlizkiy, A. D., Adekenov, S. M., Kupriyanov, A. N. *Sesquiterpene lactones of plants of Central Kazakhstan* (Almaty: Science, 1987), 240 pp.

(Кагарлицкий, А. Д., Адекенов, С. М., Куприянов, А. Н. *Сесквитерпеновые лактоны растений Центрального Казахстана* (Алматы: Наука, 1987), 240 с.).

Kalachaveedu, M., Raghavan, D., Telapolu, S., Kuruvilla, S., Kedike, B. (2018). "Data set on the characterization of the phytoestrogenic extract and isolated compounds of the roots of *Inula racemosa* Hook F (Asteraceae)." *Data in Brief*, 17: 770-773.

Karamenderes, C., Zeybek, U. (2000). "Composition of the essential oils of *Inula viscosa*, *Inula graveolens* and *Inula helenium* subsp. *turcoracemosa.*" *Journal of Faculty of Pharmacy of Ankara University*. 33: 3-5.

Karan, T., Yildiz, I., Aydin, A. Erenler, R. (2018). "Inhibition of Various Cancer Cells Proliferation of Bornyl Acetate and Essential Oil from *Inula graveolens* (Linnaeus) Desf." *Records of Natural Products*. 12 (3): 273-283.

Kaur, M., Saraf, I., Kumar, R., Singh, I. P., Kaur, S. *Bioefficacy of Hexane Extract of Inula racemosa (Asteraceae) Against Spodoptera litura (Lepidoptera: Noctuidae)*. (Deutschland: Springer-Verlag GmbH, 2018; Springer Nature (1 part), 2019). https://doi.org/10.1007/s10343-019-00462-w.

Khalil, E. A., Afifi, F. U., Al-Hussaini M. (2007). "Evaluation of the wound healing effect of some Jordanian traditional medicinal plants formulated in Pluronic F127 using mice (*Mus musculus*)." *J. Ethnopharmacol*., 109 (1): 104-112. https://doi.org/10.1016/j.jep.2006.07.010.

Kheyar-Kraouche, N., Silva, A. B. D., Serra, A. T, Bedjou, F., Bronze, M. R. (2018). "Characterization by liquid chromatography–mass spectrometry and antioxidant activity of an ethanolic extract of *Inula viscosa* leave." *Journal of Pharmaceutical and Biomedical Analysis*, 156: 297–306.

Kiliç, O. (2014). *"Chemical Composition of Two Inula sp. (Asteraceae) Species from Turkey.*" In *Iğdır Üni. Fen Bilimleri Enst. Der. Iğdır Univ.*, edited by Araştırma Makalesi. Turkey, 2014. *J. Inst. Sci. & Tech.,* 4 (1): 15-19.

Kiliç, O., Baggi, E. (2013). "Chemical Composition of Endemic *Inula macrocephala* Boiss. and Kotschy ex Boiss. from Turkey." *Asian Journal of Chemistry*. 25 (14): 7952-7954.

Krizman, V. A., Stanzo, V. V. "*Rubbers and elastomers*" In *Encyclopedic Dictionary of a Young Chemist*, 104-107. Moscow: Pedagogy, 2nd ed., 1990 (Крицман, В. А., Станцо, В. В. "*Каучуки и эластомеры*" In *Энциклопедический словарь юного химика*, 104-107. Москва: Педагогика, 2-е изд., 1990).

Kulyyasov, A. T., Edil'baeva, T. T., Turdybekov, K. M., Raldugin V. A., Shakirov M. M., Adekenov S. M. (1999). Chemical transformations of ajanolide A." *Chemistry of Natural Compounds*, 35: 55-60. https://doi.org/10.1007/BF02238210.

Kulyyasov, A.T., Seitembetov, T.S., Turdybekov, K.M., Adekenov S. M. (1996). "Epoxidation of alantolactone and isoalantolactone." *Chemistry of Natural Compounds*, 32: 869-872. https://doi.org/10.1007/BF01374017.

Kumar, C., Kumar, A., Nalli, Y., Lone, W. I., Satti, N.K., Verma, M. K., Ahmed, Z., Ali, A. (2019). "Design, synthesis and biological evaluation of alantolactone derivatives as potential anti-inflammatory agents." *Medicinal Chemistry Research*, 28: 849-856. https://doi.org/10.1007/s00044-019-02337-1.

Laurentis, Onicolino de, Losacco, V., Milillo, M. A, Limpia, Lai. (2002). "Chemical investigations of volatile constituents of *Inula viscosa* (L.) Aiton (Asteraceae) from different areas of Apulia, Southern Italy." *Delpinoa*. Italia. 44: 115-119.

Lei, J. C., Yu, J. Q., Yin, Y., Liu, Y. W., Zou, G. L. (2012). "Alantolactone induces activation of apoptosis in human hepatoma cells." *Food and Chemical Toxicology*, 50: 3313-3319.

Li, X., Huang, M., Liu, F., Gao, S., Bai, J., Liu, S., Zhou, B., Guo, Y., Zhou, H. (2020). "Analysis of chemical composition of *Inula japonica* Thunb. extract and *in vitro* screening for antipulmonary fibrosis active components." *Phytochemistry Letters*, 36: 144-149.

Li, Z., Qin, B., Qi, X., Mao, J., Wu, D. (2016). Isoalantolactone induces apoptosis in human breast cancer cells via ROS-mediated mitochondrial pathway and downregulation of SIRT1." *Arch. Pharm. Res.*, 39: 1441-1453.

Liu, B., Han, M., Sun, R. H., Wang, J. J., Liu, Y. P., Wen J. K. (2011). "Acetylbritannilactone induces G1 arrest and apoptosis in vascular smooth muscle cells." *International Journal of Cardiology*, 149: 30-38.

Liu, C. H., Mishra, A. K., Tan, R. X. (2006). "Repellent, insecticidal and phytotoxic activities of isoalantolactone from *Inula racemosa.*" *Crop Protection*, 25 (5): 508-511.

Lounis, H., Bergheim, I., Bouhaimi, A., Guigonis, J. M., Belhamel, K. (2018). "Anti-inflammatory and antioxidant activities of *Inula viscosa* and *Senecio anteuphorbium.*" *Oriental Pharmacy and Experimental Medicine*, 18: 225-236.

Lovinskaya, A. V., Kolumbayeva, S. Zh., Shalakhmetova, T. M., Marsova, M. V., and Abilev, S. K. (2017). "Antigenotoxic Activity of Biologically Active Substances from *Inula britannica* and *Limonium gmelini.*" *Russian Journal of Genetics*, 53 (12): 1311-1319.

Lu, Y., Li, Y., Jin, M., Yang, J. H., Li, X., Chao, G. H., Park, H. H., Parka, Y. N., Son, J. K., Lee, E., Chang, H. W. (2012). "*Inula japonica* extract inhibits mast cell-mediated allergic reaction and mast cell activation." *Journal of Ethnopharmacology*, 143: 151-157.

Ma, Y. Y., Zhao, D. G., Gao, K. (2013). "Structural Investigation and Biological Activity of Sesquiterpene Lactones from the Traditional Chinese Herb *Inula racemosa*." *Journal Natural Product*, 76 (4): 564-570.

Macias, F. A., Santana, F. A., Durán, A. G., Peralta, A. C., Galindo, J. C., Molinillo, J. M. G. "*Guaianolides for multipurpose molecular design*". In *Pest management with natural products*, *ACS Symposium Series,* Vol. 1141, Chapter 12: 167-188. New York, USA, 2013. https://doi.org/10.1021/bk-2013-1141.ch012.

Makhmudov, M. K., Plutno, A. B., Sham'yanov, I. D., Tashkhodzhaev B. (1995). "Stereochemistry of the products of the hydrogenation of leucomisin." *Chemistry of Natural Compounds*, 31: 211-213 https://doi.org/10.1007/BF01170208.

Makhmudov, M. K., Tashkhodzhaev, B. & Zakirov, S. K. (1990b). "Molecular and crystal structure of the sesquiterpene lactone ketopelenolide A." *Chemistry of Natural Compounds*, 26: 283-285. https://doi.org/10.1007/BF00597849.

Makhmudov, M. K., Tashkhodzhaev, B., Abduazimov, B. K. (1993). "Spatial structure of the dimeric sesquiterpene germacrane lactone mikagoyanolide." *Chemistry of Natural Compounds*, 29: 174-178. https://doi.org/10.1007/BF00630109.

Makhmudov, M. K., Tashkhodzhaev, B., Saidkhodzhaev, A. I, Ibragimov B. T. (1990a). "X-ray structural investigation of sesquiterpene esters of plants of the genus *Ferula*. VI. Crystal and molecular structure of ugaferin." *Chemistry of Natural Compounds*, 26: 157-160 https://doi.org/10.1007/BF00607533.

Makhmudov, M. K., Tashkhodzhaev, B., Zhonkhozhaeva, F. B, Sham'yanov I. D. "Pseudoguaianolides of *Inula aspera*: Spatial structure of erigerolide." *Chemistry of Natural Compounds*, 29: 191–195 (1993b). https://doi.org/10.1007/BF00630113.

Manez, S., Hernandez, V., Giner, R. M., Rios, J. L., Recio, M. del C. (2007). "Inhibition of pro-inflammatory enzymes by inuviscolide, a sesquiterpene lactone from *Inula viscosa.*" *Fitoterapia*, 78: 329-331.

Matejić, J., Šarac, Z. and Ranđelović, V. (2010). "Pharmacological Activity of Sesquiterpene Lactones." *Biotechnology & Biotechnological Equipment*, 24 (1): 95-100. http://dx.doi.org/10.1080/13102818.2010.10817819.

Michalakea, E., Graikou, K., Aligiannis, N., Panoutsopoulos, G., Kalpoutzakis, E., Roussakis, C., Chinou, I. (2019). "Isolation and structure elucidation of secondary metabolites of two Greek endemic *Inula* species: Biological activities." *Phytochemistry Letters*, 31: 155-160.

Mitić, V. D., Ilic M., Stankov-Jovanovic V., Djordjevic, A. S., Marković, M., Stojanović, G. S. (2019). "Volatiles composition and antioxidant activity

Inula oculus-christi L. from Serbia." *Natural Product Research.* http://dx.doi.org/10.1080/14786419.2018.155076.

Mohan, S., Gupta, D. (2017). "Phytochemical analysis and differential *in vitro* cytotoxicity assessment of root extracts of *Inula racemose.*" *Biomedicine & Pharmacotherapy*, 89: 781-795.

Mukhamatkhanova, R. F. "*Secondary metabolites of subgenus Seriphidium of genus Artemisia plants.*" Diss. PhD in Chem. Sciences (Tashkent, 2018), 149 pp. (Мухаматханова, Р. Ф. "Вторичные метаболиты растений рода *Artemisia* подрода *Seriphidium.*" Дисс. PhD по хим. наукам (Ташкент, 2018), 149 с).

Nam, K. W., Oh, G. T., Seo, E. K., Kim, K. H., Koo, U., Lee, S. J. (2009). "Woongchon Mar. Nuclear factor kB-mediated down-regulation of adhesion molecules: Possible mechanism for inhibitory activity of bigelovin against inflammatory monocytes adhesion to endothelial cells." *Journal of Ethnopharmacology*, 123: 250-256.

Nasser, M., Housheh, S., Kourini, A., Maala, N. (2014). "Chemical composition of essential oil from leaves and flowers of *Inula viscosa* (L.) in Al-Qadmus region, Syria." *IJPSR* 5 (12): 5177-5182.

Nie, L. Y., Qin, J. J., Huang, Y., Yan, L., Liu, Y. B., Pan, Y. X., Jin, H. Z., Zhang, W. D., (2010). "Sesquiterpenoids from *Inula leariifolia* Inhibit Nitric Oxide Production." *Journal Natural Product*, 73 (6): 1117-1120. https://doi.org/10.1021/np100124a.

Nikonova, L. P., Nikonov, G. K., (1970). "Igalan, a new sesquiterpene lactone from Inula grandis." *Chemistry of Natural Compounds*, (6): 528-531. https://doi.org/10.1007/BF00563431.

Omezzinea, F., Ladhari, A., Rinez, A., Haouala, R. (2011). "Potent herbicidal activity of *Inula crithmoides* L." *Scientia Horticulturae*, 130: 853-861.

Paliwal, S. K., Sati, Bh., Faujdar, S., Sharma, S. (2017). "Antioxidant and antibacterial activities of various extracts of *Inula cuspidata* C.B. Clarke stem." *Beni-Suef University Journal of Basic and Applied Sciences*, 6 (2): 97-105.

Pan, M. H.; Chiou, Y.; Cheng, A. C.; Bai, N.; Lo, C. Y.; Tan, D.; Ho, C. T. (2007). "Involvement of MAPK, Bcl-2 family, cytochrome-C, and caspases in induction of apoptosis by 1,6-*O,O*-diacetylbritannilactone in human leukemia cells." *Mol. Nut. Food Res.*, 51: 229-238.

Park, E. J., Kim, Y. M., Park, S. W., Kim, H. J., Lee, J. H., Lee, D. U., Chang, K. C. (2013). "Induction of HO-1 through p38 MAPK/Nrf2 signaling pathway by ethanol extract of *Inula helenium* L. reduces inflammation in LPS-activated RAW 264.7 cells and CLP-induced septic mice." *Food and Chemical Toxicology*, 55: 386–395.

Park, E. H., Bae, W. Y., Kim, J. Y., Kim, K. T., Paik, H. D. (2017). "Antimelanogenic effects of *Inula britannica* flower petal extract fermented by *Lactobacillus plantarum* KCCM 11613P." *J. Zhejiang Univ.-Sci. Biomed. & Biotechnol.*, 18 (9): 816-824.

Park, H. H., Kim, M. J., Li, Y., Park, Y. N., Lee, J., Lee, Y. J., Kim, S. G., Park, H. J., Son, J. K., Chang, H. W., Lee, E. K. (2013). "Britanin suppresses LPS-induced nitric oxide, PGE2 and cytokine production via NF-κB and MAPK inactivation in RAW 264.7 cells." *International Immunopharmacology*, 15: 296-302.

Piet, D. P., Schrijvers, R., Frassen, M. C. R., De Groot, A. (1995). "Biotransformation of germacrene epoxides by *Chichorium intybus* L." *Tetrahedron*, 51 (22): 6303-6314.

Plant resources of Russia and neighboring countries: Flowering plants, their chemical composition and use, edited by Budanzev A. L., 9: 575 pp. St. Petersburg: Peace and family, 1996.

Plant resources: Flowering plants, their chemical composition, use, edited by Sokolov P. D., 7: 352 pp. St. Petersburg: Science, 1993.

Plemenkov, V. V. *Introduction to the Chemistry of Natural Compounds* (Kazan, 2001), 376 pp. (Племенков, В. В. *Введение в химию природных соединений* (Казань, 2001), 376 с.).

Ponomarev, D. L., Turdybekov, K. M., Buketova, G. K, Turmukhambetov A. Zh., Adekenov S. M. (1997). "Spatial structure of isoepoxyestafiatin." *Chemistry of Natural Compounds*, 33: 50-51. https://doi.org/10.1007/BF02273922.

Priydarshi, R., Melkani, Anand B., Mohan, L., Pant Charu, C. (2015). "Terpenoid composition and antibacterial activity of the essential oil from *Inula caPa* (Buch-Ham. ex. D. Don) DC." *Journal of Essential oil research.* 2 p. http://dx.doi.org/10.1080/10412905.2015.1090935.

Pukhova, S. A., Afanasyeva, S. V., Anikina, L. V., Semakov, A. V., Dubrovskaya, E. S., Klochkov, S. G. (2018). "Amino Derivatives of Natural Epoxyalantolactone: Synthesis and Cytotoxicity toward Tumor Cells." *Russian Journal of Bioorganic Chemistry*, 44 (5): 553-561.

Qin J. J., Zhu J. X., Zeng Q., Cheng X. R., Zhang S. D., Jin H. Z., Zhang W. D. (2012). "Sesquiterpene lactones from *Inula hupehensis* inhibit nitric oxide production in RAW264.7 macrophages." *Planta Medica*, 78: 1002-1009.

Qin, J. J., Jin, H. Z., Fu, J. J., Hu, X. J., Wang, Y., Yan, S. K., Zhang, W. D. (2009). "Japonicones A–D, bioactive dimeric sesquiterpenes from *Inula japonica* Thunb." *Bioorganic & Medicinal Chemistry Letters*, 19: 710-713.

Qin, J. J., Jin, H. Z., Huang, Y., Zhang, S. D., Shan, L., Voruganti, S., Nag, S., Wang, W., Zhang, W. D., Zhang, R. (2013). "Selective cytotoxicity, inhibition of cell cycle progression, and induction of apoptosis in human breast cancer cells by sesquiterpenoids from *Inula lineariifolia* Turcz." *European Journal of Medicinal Chemistry*, 68: 473-481.

Qin, J. J., Jin, H. Z., Zhu, J. X., Fu, J. J., Zeng, Q., Cheng, X. R., Zhu, Y., Shan, L., Zhang, S. D., Pan, Y. X., Zhang, W. D. (2010). "New sesquiterpenes from *Inula japonica* Thunb. with their inhibitory activities against LPS-induced NO production in RAW264.7 macrophages." *Tetrahedron*, 66 (48): 9379-9388. https://doi.org/10.1016/j.tet.2010.09.091.

Rasul, A., Khan, M., Yu, B., Ali, M., Bo, Y. J., Yang, H., Ma, T. (2013). "Isoalantolactone, a sesquiterpene lactone, induces apoptosis in SGC-7901 cells via mitochondrial and phosphatidylinositol 3-kinase/Akt signaling pathways." *Archives of Pharmacol Research*, 36 (10): 1262-1269. https://doi.org/10.1007/s12272-013-0217-0.

Ren, J., Qin, J. J., Cheng, X. R., Yan, S. K., Jin, H. Z., Zhang, W. D. (2013). "Five new sesquiterpene lactones from *Inula hupehensis.*" *Archives of Pharmacol Research*. - Korea, 36: 1319-1325. https://doi.org/10.1007/s12272-013-0135-1.

Rethyl, B., Csupor-Loffler, B., Zupko, I., Hajdu, Z., Mathe, I., Hohmann, J., Redei, T. Falkay, G. (2007). "Antiproliferative Activity of Hungarian Asteraceae Species against Human Cancer Cell Lines.": Part I. *Phytotherapy Research*, 21: 1200-1208.

Rozenblat, S., Grossman, S., Bergman, M., Gottlieb, H., Cohen, Y., Dovrat, S. (2008). "Induction of G2/M arrest and apoptosis by sesquiterpene lactones in human melanoma cell lines." *Biochemical pharmacology*, 75: 369-382.

Rybalko K. S. *Natural sesquiterpene lactones* (Moscow: Medicine, 1978), 320 pp. (Рыбалко К. С. *Природные сесквитерпеновые лактоны* (Москва: Медицина, 1978), 320 с.).

Seca, Ana M. L., Pinto, Diana C. G. A., Silva, Artur M. S. (2015). "Metabolomic Profile of the Genus *Inula.*" *Chemistry & Biodiversity,* 12 (6): 859-906.

Semakov, A. V., Anikina, L. V., Afanasyeva, S. V., Pukhov, S. A., Klochkov, S. G. (2018). "Synthesis and Antiproliferative Activity of Conjugates of Anthracycline Antibiotics with Sesquiterpene Lactones of the Elecampane." *Russian Journal of Bioorganic Chemistry*, 44 (5): 538-546.

Semenov, A. A. *Essay on the Chemistry of Natural Compounds* (Novosibirsk: Nauka, 2000), 77-161 (Семенов, А. А. *Очерк химии природных соединений* (Новосибирск: Наука, 2000), 77-161).

Sham'yanov, I. D., Tashkhodzhaev, B., Mukhamatkhanova, R. F., Antipin M. Yu. (2000a). "9β-hydroxy-11β(H),13-dihyhdrosantamarine, a new eudesmanolide from *Artemisia leucotricha.*" *Chemistry of Natural Compounds*, 36: 284-287. https://doi.org/10.1007/BF02238337.

Sham'yanov, I. D., Tashkhodzhaev, B., Mukhamatkhanova, R. F., Antipin M. Yu. (2000b). "1α-Acetoxy-11β(H),13-dihydrodouglanin, a New Eudesmanolide from *Artemisia lehmanniana.*" *Chemistry of Natural Compounds*, 36: 369-372. https://doi.org/10.1023/A:1002888713138.

Sowndhararajan, K., Choa, H., Yu, B., Song, J., Kim, S. (2016). "Effect of inhalation of essential oil from *Inula helenium* L. root on electroencephalographic (EEG) activity of the human brain." *European Journal of Integrative Medicine*, 8: 453-457.

Stojanović-Radić, Z., Čomić, Lj., Radulović, N., Blagojević, P., Denić, M., Miltojević, A., Rajković, J., Mihajilov-Krstev, T. (2012). "Antistaphylococcal activity of *Inula helenium* L. root essential oil: eudesmane sesquiterpene lactones induce cell membrane damage." *European Journal of Clinical Microbiology and Infectious Diseases.* 31: 1015-1025. https://doi.org/10.1007/s10096-011-1400-1.

Su, B. N., Takaishi, Y., Tori, M., Takaoka, S., Honda, G., Itoh, M., Takeda, Y., Kodzhimatov, O. K. and Ashurmetov, O. (2000). "Macrophyllidimers A and B, two novel sesquiterpene dimers from the bark of *Inula macrophylla.*" *Tetrahedron Letters*, 41: 1475-1479.

Su, R. X., Wu, Z. L., Wang, Y. Y., Dong, H. L., Dong, H. Y., Li, H. L., Wang, S. M. (2020) "Leariifolianoids M-O, three rare sesquiterpene lactone dimers inhibiting NO production from *Inula lineariifolia.*" *Fitoterapia*, 141: 104454.

Talzhanov, N. A., Rodichev, M. A., Raldugin, V. A., Gatilov Yu. V., Shakirov M. M., Ivasenko S. A., Turdybekov K. M., Adekenov S. M. (2007). "Tourneforin, a novel eudesmanolide from *Artemisia tournefortiana.*" *Chemistry of Natural Compounds*, 43: 555-557. https://doi.org/10.1007/s10600-007-0190-2.

Tang, J. J., He, Q. R., Dong, S., Guo, X., Wang, Y. G., Lei, B. L., Tian, J. M., Gao, J. M. (2018). "Diversity Modification and Structure-Activity Relationships of Two Natural Products 1β-hydroxy Alantolactone and Ivangustin as Potent Cytotoxic Agents." *Sci. Rep.*, 8 (Jan): 1722. PMID 29379131. https://doi.org/10.1038/s41598-018-20192-9).

Te, Z., Fei, G., Lin, Z., Tian-You, S. (2013). "Essential Oil from *Inula britannica* Extraction with SF-CO_2 and Its Antifungal Activity." *Journal of Integrative Agriculture*, 12 (10): 1791-1798.

Tleugabulov, N. S., Turdybekov, K. M., Raldugin, V. A., Shakirov M. M., Kulyyasov A. T., Adekenov S. M. (1998). "Spatial structures of pulchellin C and alantolactone epoxides." *Chemistry of Natural Compounds*, 34: 462-465. https://doi.org/10.1007/BF02329595.

Todorova, M., Trendafilova, A., Dimitrov, D. (2017). "Chemical composition of volatile extract from *Inula aschersoniana* Janka var. *aschersoniana* growing in Bulgaria." *Lekovite Sirovine*, Serbia, 37: 21-24.

Todorova, M., Trendafilova, A., Ivanova, V., Danova, K., Dimitrov, D. (2017). "Essential oil composition of *Inula britannica* L. from Bulgaria." *Natural*

Product Research, 31 (14): 1693-1696. https://doi.org/10.1080/14786419.2017.1285295.

Trendafilova, A., Ivanova, V., Todorova, M., Aneva, I. (2017). "New sesquiterpene lactones from *Inula oculus-christi* L." *Phytochemistry Letters*, 21: 221-225.

Turdybekov, K. M., Edil'baeva, T. T., Raldugin, V. A., Shakirov M. M., Kulyyasov A. T., Adekenov S. M. (1998a). "Structure of subchrysin." *Chemistry of Natural Compounds*, 34: 141-144. https://doi.org/10.1007/BF02249130.

Turdybekov, K. M., Edil'baeva, T. T., Raldugin, V. A., Shakirov M. M., Kulyyasov A. T., Adekenov S. M. (1998b). "Molecular and crystal structures of argolide epoxide." *Chemistry of Natural Compounds*, 34: 44-47. https://doi.org/10.1007/BF02249684.

Turdybekov, K. M., Turmukhambetov, A. Z., Adekenov, S. M., Struchkov Yu. T. (1994). "Anolide - A new guaianolide from *Achillea nobilis*." *Chemistry of Natural Compounds*, 30: 460-463. https://doi.org/10.1007/BF00630399.

Turmukhambetov, A. Z., Adekenov, S. M., Turdybekov, K. M., Gatilov Yu. V. (1991). "Gracilin - A new sesquiterpene lactone from *Artemisia gracilescens*." *Chemistry of Natural Compounds*, 27: 292-295 https://doi.org/10.1007/BF00630312.

Vasas, A., Hohmann, J. (2011). "Xanthane sesquiterpenoids: structure, synthesis and biological activity." *Natural Product Reports*, 28: 824.

Venkateswara, Rao G., Vijaya, Bhaskar Reddy M., Pirabakaran, R., Madhavi, M. S. L., Kavitha, K. and Mukhopadhyay, T. (2013). "Antimicrobial and melanin synthesis inhibitory activities of the roots of *Inula racemosa* Hook f." *Archives of Applied Science Research*, 5 (2): 104-108.

Verma, R. S., Padalia, R. C., Chauhan, A. (2014). "Leaf essential oil composition of *Inula cuspidata* (Wall. ex DC.) C. B. Clarke from India." *The Journal of essential oil research,* 26 (4): 233-237.

Vvedenskiy, A. I. *Flora of Uzbekistan* (Tashkent: Publishing House of the Academy of Sciences of the Uzbek SSR, 1962), 6: 562-631. (Введенский, А. И. *Флора Узбекистана* (Ташкент: Издательство Академии Наук УзССР, 1962), 6: 562-631).

Wang, F. Y., Li, X. Q., Sun, Q., Yao, S., Ke, C. Q., Tang, C. P., Liu, H. C., Geng, M. Y., Ye, Y. (2012). "Sesquiterpene lactones from *Inula cappa.*" *Phytochemistry Letters*, (5): 639–642.

Wang, M., Wang, K., Gao, X., Zhao, K., Chen, H., Xu, M. (2018). "Anti-inflammatory effects of isoalantolactone on LPS-stimulated BV2 microglia cells through activating GSK-3β-Nrf2 signaling pathway." *International Immunopharmacology*, 65: 323-327.

Wang, X., Tang, Sh. An, Wang, R., Qiu, Y., Jin, M. and Kong, D. (2015). "Inhibitory Effects of JEUD-38, a New Sesquiterpene Lactone from *Inula japonica* Thunb, on LPS-Induced iNOS Expression in RAW264.7 Cells." *Inflammation*, 38 (3): 941-948. https://doi.org/10.1007/s10753-014-0056-2.

Wang, X., Yu, Z., Wang, C., Cheng, W., Tian, X., Huo, X., Wang, Y., Sun, C., Feng, L., Xing, J., Lan, Y., Sun, D., Hou, Q., Zhang, B., Ma, X., Zhang, B. (2017). "Alantolactone, a natural sesquiterpene lactone, has potent antitumor activity against glioblastoma by targeting IKKβ kinase activity and interrupting NF-κB/COX-2-mediated signaling cascades." *Journal of Experimental & Clinical Cancer Research,* 36: 93. https://doi.org/10.1186/s13046-017-0563-8.

Wang, Z., Geng, L., Chen, Z., Lin, B., Zhang, M., Zheng, S. (2017). "*In vivo* therapeutic potential of *Inula racemosa* in hepatic ischemia-reperfusion injury following orthotopic liver transplantation in male albino rats." Wang et al. *AMB Expr.*, 7: 211. https://doi.org/10.1186/s13568-017-0511-1.

Wen, S. W., Zhang, Y. F., Li, Y., Xu, Y. Z., Li, Z. H., Lu, H., Zhu, Y. G., Liu, Z. X., Tian, Z. Q. (2018). "Isoalantolactone Inhibits Esophageal Squamous Cell Carcinoma Growth Through Downregulation of MicroRNA-21 and Derepression of PDCD4." *Digestive Diseases and Sciences*, 63: 2285-2293.

Wu, J. W., Tang, C. P., Cai, Y. Y., Ke, C. Q., Lin, L. G., Yao, S., Ye, Y. (2017). "Cytotoxic germacrane-type sesquiterpene lactones from the whole plant of *Inula cappa." Chinese Chemical Letters*, 28 (5): 927-930.

Wu, X. D., Ding, L. F., Liu, K. X., Zhang, B.-J., Xin, X. L., Huang, J. (2008). "Structural determination of three new eudesmanolides from *Inula helenium." Magnetic resonance in chemistry*, 46: 1084.

Wu, X. D., Ding, L. F., Tu, W. C., Yang, H., Su, J., Peng, L. Y., Li, Y., Zhao, Q. S. (2016). "Bioactive sesquiterpenoids from the flowers of *Inula japonica." Phytochemistry*, 129: 68-76.

Xu, L.-W., Shi, Y.-P. (2011). "Sesquiterpenoids from *Inula racemosa." Journal of Asian Natural Products Research*, 13: 570-574.

Xu, R., Peng, Y., Wang, M., Li, X. (2019). "Intestinal Absorption of Isoalantolactone and Alantolactone, Two Sesquiterpene Lactones from Radix Inulae, Using Caco-2 Cells." *European Journal of Drug Metabolism and Pharmacokinetics*, 44: 295-303.

Xu, X. Y., Sun, P., Guo, D. A., Liu, X., Liu, J. H., Hu, L. H. (2015). "Cytotoxic sesquiterpene lactone dimers isolated from *Inula japonica*." *Fitoterapia*, 101 (March): 218-223.

Yang, J. L., Wang, R., Liu, L. L., Shi, Y. P. (2011). "Sesquiterpenoids from *Inula britannica*." *Planta Medica*, 77 (4): 362-367. https://doi.org/10.1055/s-0030-1250345.

Yu, Z. P., Zhang, J. S., Zhang, Q., Yu, S. J., Zhang, Y., Yu, J. H., Zhang, H. (2019). "Bioactive sesquiterpenoids and sesquiterpenoid glucosides from the flowers of *Inula japonica*." *Fitoterapia*, 138: 104292.

Yue, G. G. L., Chan B. C. L., Kwok H. F., Wong Y. L., Leung H. W., Ji C. J., Fung K. P., Leung P. Ch., Tan N. H., Lau C. B. S. (2013). "Anti-angiogenesis and immunomodulatory activities of an anti-tumorsesquiterpene bigelovin isolated from *Inula helianthus-aquatica*." *European Journal of Medicinal Chemistry*, 59: 243-252.

Zangeneh, M. Mahdi, Zangeneh A., Almasi M., Tahvilian R., Hosseini F., Morad R. (2018). "A comparative study of hepatoprotective effect of *Inula britannica* L aqueous extract and glibenclamide in streptozotocin-induced diabetic mice." *Comparative Clinical Pathology,* 27: 1649–1657.

Zeggwagh, N. A., Ouahidi, M. L., Lemhadri, A., Eddouks, M. (2006). "Study of hypoglycaemic and hypolipidemic effects of *Inula viscosa* L. aqueous extract in normal and diabetic rats." *Journal of Ethnopharmacology*, 108 (2): 223-227. https://doi.org/10.1016/j.jep.2006.05.005

Zhang, T., Gong, T., Yang, Y., Chen, R. Y., Yu, D. Q. (2012). "Two new eudesmanolides from *Inula racemosa* and their bioactivities." *Phytochemistry Letters*, 5: 229-232.

Zhang, X. F., Du, J. L., Ren, J., Ye, F. M., Xie, Y. G., Cheng, X. R., Yan, S. K., Jin, H. Z. (2015). "Three new sesquiterpene lactones from *Inula britannica.*" *Archives of Pharmacal Research*, 38: 666-672. https://doi.org/10.1007/s12272-014-0388-3.

Zhao, Y., Tao, M., Wang, R., Guo, Y., Wang, M. (2020). "Japonicone V, a sesquiterpene lactone derivative from the flowers of *Inula japonica*, inhibits hepatitis E virus replication by targeting virus-associated autophagy." *Journal of Functional Foods*, 65: 103755.

Zheng, B., Wu, L., Ma, L., Liu, S., Li, L., Xie, W., Li, X. (2013). "Telekin Induces Apoptosis Associated with the Mitochondria-Mediated Pathway in Human Hepatocellular Carcinoma Cells." *Biological and Pharmaceutical Bulletin*, 36 (7): 1118-1125.

Zhu, S. L., Huang, C. L., Xiao, Y. Z., Xie, Y. G., Wu, G. J., Zhang, W. D., Zhang, Y., Chen, D. J., Jin, H. Z. (2019). "Pterocaullins A–D, four sesquiterpene lactones from *Inula pterocaula*." *Phytochemistry Letters*, 33: 70-76.

Zhu, Sh., Xiao, Y., Wu G., Xie, Y., Muhammad, I., Sun, Yu., Yan, Sh. K., Jin, H. Z. (2019). "Chemical constituents of *Inula pterocaula.*" *Chemistry of Natural Compounds*, 55 (6): 1135-1137. https://doi.org/10.1007/s10600-019-02913-2.

Chapter 2

Phytochemical Prospection of Biologically Active Terpenoids from Saharan Endemic Medicinal Plants

Abdelkrim Cheriti*
Phytochemistry and Organic Synthesis Laboratory
Faculty of Medicine, UTMB, Algeria

Dedicated to my mother Dahmani Lalia, on the occasion of her 78th birthday.

Abstract

The Algerian Sahara is an interesting area of biodiversity, which is a source of rich and abundant medical matter and an important reservoir for the research of new bioactive compounds with pharmacological potential. In Sahara many species from Asteraceae family are used in folk medicine for the treatment of diseases. Several studies demonstrated the antimicrobial, anti-inflammatory, antioxidant, anti-hyperlipidemia, hepatoprotection, cytoprotective, giving protection against cardiovascular disease and certain forms of cancer of Asteraceae species which contained a diversity of chemical compounds especially terpenoids, sesquiterpene lactones, triterpenoid saponin, steroids and polyphenolic compounds. Among these compounds, terpenoids are

*Corresponding Author's E-mail: karimcheriti@yahoo.com.

In: Terpenes and Terpenoids
Editor: Charles A. Davies
ISBN: 978-1-68507-559-0

chemio-characteristic of Asteraceae family and having a potential application in pharmaceutical and cosmetic industries.

In this chapter we cover the nature of terpenoids together with their biological activities and their isolation from endemic species from Asteraceae family growing in Algerian Sahara and used as medicinal plants, namely: *Asteriscus graveolens*, *Launaeaarborescensand Warionia saharae*

Keywords: terpenoids, asteraceae, endemic, phytochemical, analysis, biological activities, Sahara, Algeria

1. Introduction

Natural products from plants used in traditional pharmacopeia have been employed since ancient time for the treatment and management of various ailments around the world and have been successfully applied in clinics and as an important source of new compounds for drugs discovery.

Thus, one of the most efficient ways of finding new drugs from natural compounds is collecting data on the use of medicinal plants as recommended by the World Health Organization (WHO) which has recognized the potential utility of traditional remedies and has listed 21000 plants which are used for medicinal purposes around the Globe [1-3].

It is well known that, nearly 50,000 species of higher plants have been used for medicinal purposes and natural products with their derivatives represent more than 50% of all the drugs in clinical use and in which higher plants contribute to no less than 25% [2]. So, the isolation, identification and synthesis of natural molecules continue to plays a central role in chemistry; pharmacology and related disciplines.

A rich chemical diversity is found in flora species which can contain up to several thousands of different bioactives molecules, among them, terpenoids known since old civilizations and cultures by using essential oils for many purposes. Actually, these natural products have become of widespread importance in pharmacology, perfumery, beverages, foods and detergents [2].

Structurally, terpenoids comprise the largest class of natural products (60% of known natural products) displaying a wide range of biological and pharmacological activities, due to their chemical structure diversity with so many functions that originate from squalene or oxidosqualene by a series of intramolecular condensation reactions and can have a simple aliphatic structure, or a cyclic one. Monocyclic, bicyclic, tricyclic and polycyclic structures [4]. Thus, the different therapeutic activities and the diversity of structures shown by this class of natural compounds make the isolation from plants, identification, synthesis and biosynthesis of new natural terpenoids an interesting research field.

The richness of Algerian flora about 4450 taxa of which 3950 indigenous with 6.5% endemic due to the climate diversity[5], allows the country to have a source of rich and abundant medical matter, which necessary contribute to lead several programmes for sustainable development.. In particular, Algerian Sahara constitutes an important reservoir of many plants species which have not been investigated until today. Among this flora, some endemic species from Asteraceae family have been widely used in the Algerian Sahara ethnopharmacopea for the treatment of various diseases such as: *Asteriscus graveolens, Launaeaarborescens and Warionia saharae* [6-9].

We present in this chapter the importance of these three endemic medicinal species and the chemical nature and biological activities of terpenoids isolated from these Saharan endemic medicinal species belonging to Asteraceae family.

2. Taxonomy and Botanical Description of the Endemic Asteraceae Species

The science of natural products is a very broad one, ranging from taxonomy to chemistry, biology and pharmacology. The isolation and structure elucidation of complex chemical structures are linked to their biology and biosynthesis importance [10, 11]. Methodologically, research on natural products obtained from medicinal plants necessarily involves ethnopharmacological and taxonomic studies for the selection and identification of species before starting the phytochemistry and biological works.

Algeria has a rich biodiversity flora about 3139 species with 1000 medicinal species, 700 endemic, 315 rather scarce, 590 scarce species, 730 very scarce, 35 extremely scarce and 226 species are threatened with extinction [11, 12].. In another hand, much of the Algerian territory is covered by the Saharan desert and the Algerian Sahara is one of the warmest areas in the world [16, 17]. The flora of the Algerian Sahara has a potential medicinal flora very rich and diversified with a very pronounced endemism especially. It includes more than 960 species to be organized following different types of habitats or landscapes and consists of several formations, namely: vegetation of ergs (or desert dunes); vegetation of regs (or stone desert) and the vegetation of salty soils [12, 13]. According to Ozenda [13], Asteraceae family represents 13.8%, 11.2% and 7.9% of the total flora respectively in Sptentrional Sahara, Central Sahara and Meridional Sahara.

From a botanical point of view, the family Asteraceae (Compositae) (Known as the aster, daisy, or sunflower family) is the largest family of flowering plants, in terms of number of species. It comprises about 1600 genera and more than 25000 species of herbaceous plants, shrubs, and trees [14]. The largest genera are *Senecio* (1500 species), *Vernonia* (1000 species), *Cousinia* (600 species), *Centaurea* (600 species). The family is classified over three subfamilies and 13 tribes: Heliantheae, Astereae, Anthemideae, Arctotideae, Inuleae, Senecioneae, Calenduleae, Eupatorieae, Vernonieae, Cynareae, Mutisieae, Liabeae and Lactuceae (Cichorieae) [15].

The following we present a brief botanical description of the three species *Asteriscus graveolens*, *Launaea arborescens and Warionia saharae* from Asteraceae family which have been widely used in the Algerian Sahara ethnopharmacopea for the treatment of various diseases as a medicinal plant.

2.1. *Asteriscus Graveolens*

The genus *Asteriscus* is a mainly Macaronesian and North African genus belonging to the Inuleae tribe of the Asteraceae family. All species have pubescent leaves, the involucre, consisting of imbricated bracts of which the outer are conspicuous and foliaceous and the inner smaller in size and largely coriaceous. The paleas of the Asteriscus species are uniformly eglandular and contain a central, longitudinal row of secretory cavities [16]. In the Algerian flora, the genus *Asteriscus*, is represented by four species: *Asteriscus aquaticus*, *Asteriscus maritimus*, *Asteriscus pygmaeus* and *Asteriscus graveolens (Forssk).*

The specie *Asteriscus graveolens* [syn. *Asteriscus cavanillesi (Caball.), Asteriscus paui*, *Bubonium graveolens* (Forssk.) Maire, *Bubonium odorum* (Schoub.), *Buphthalmum stenophyllum* (Link), *Nauplius graveolens* subsp. *Stenophyllus* (Link), *Odontospermum graveolens* and *Odontospermum odorum*(Schousb.)], its name is a combination of *asteriskos*, which means small-star and *graveolens* that means very fragrant. It is an endemic herbaceous medicinalaromatic plant mainly distributed in south-westernAlgeria and south-eastern Morocco. It is present throughout the Sahara especially in sandy clay pans, silty-sandy, stony Wadis and on rocky mountains and it blooms in spring, from March to May [13, 17, 18] (Figure 1).

- Kingdom :Plantae
- Division:Angiosperm
- Class:Eudicots
- Subclass : Asterids
- Order:Asterales
- Family:Asteraceae
- Subfamily: Asteroideae
- Tribe:Inuleae
- Sub-tribe:Inulinae
- Genus:*Asteriscus*
- Specie: *Asteriscus graveolens*
- Vernacular name*: Tafs, Tameyout*
- N° POSL Herbarium: CA00/14

Figure 1. Botanical classification of the medicinal plant *Asteriscus graveolens* - Algeria Sahara.

2.2. *Launaea arborescens*

Launaea Cass., named after naturalist and lawyer Jean Claude Mien Mordant de Launay (1750-1816), is a small genus (tribe Lactuceae or Cichorieae, subtribe Sonchinae) mainly distributed in the South Mediterranean, Africa and SW Asia. They are perennial to annual herbs, small shrubs or sub shrubs. The genus contains about 54 species, most of which are adapted to dry, saline and sandy habits [2]. In the Algerian flora, the genus *Launaea* is represented by nine species including five endemics of North Africa: *L. angustifolia*, *L. quercifolia*, and *L. cassiniana* are the endemic plants of the North Africa, with limited distribution whereas *L. acanthoclada* and *L. arborescens* are two endemic plants of the north-west of Africa. The other four species *L. nudicaulis* and *L. residifolia* sprout in Algeria and Tunisia Mediterranean Sea, whereas *L. glomerata* and *L. mucronata* grow in the Saharan Atlas [9, 12, 13].

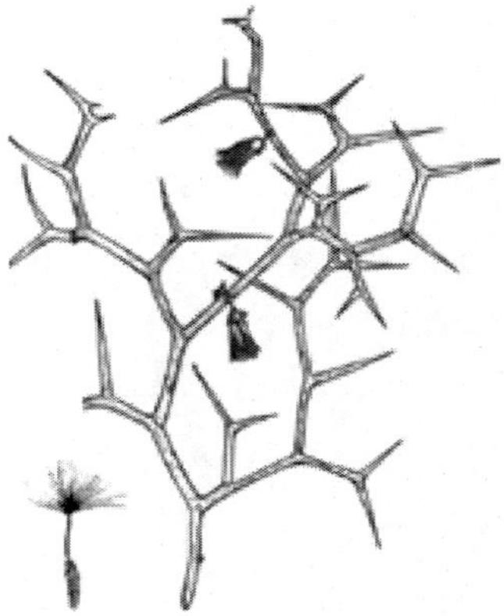

- Kingdom :Plantae
- Division:Angiosperm
- Class:Eudicots
- Subclass : Asterids
- Order:Asterales
- Family:Asteraceae
- Subfamily:Cichorioideae
- Tribe:Cichorieae
- Sub-tribe:Sonchinae
- Genus:Launeae
- Specie : *Launeae arborescens*
- Vernacular name*: Oum L'bina, Bou chlaba*
- N° POSL Herbarium: CA00/25

Figure 2. Botanical classification of the medicinal plant *Launeae arborescens* - Algeria Sahara.

Launaea arborescens (Batt.) Murb, (*arborescens, arbor*, "tree," *escens*, "becoming like"), (syn. *Zollikoferia spinosa* DC, *Zollikoferia arborescens* Batt.) is an almost leafless, xerophilous, perennial spiny shrub, 40-120 cm. high, with typical zig-zag shaped stems, growing on gravelly sand wadi beds and can survive in severe drought when the rainfall is less than 100 mm per year (Figure 2). The young stems are green, glabrous and erect. The older ones become tough spines. The leaves are narrow and dissected in small lobes, evergreen at the base but shed after flowering from the stems. The flowers are yellow, with a small involucre of 1-2 cm, with bright yellowligules, and abundant flowering occurs from March to June. The fruits are small achenes (4 x 1 mm.). The roots are very deep, the leaves and stems have white latex which is similar in appearance to milk (thus the vernacular name "*Oum loubina*") [9, 10, 12, 13, 19].

2.3. *Warionia saharae*

The genus *Warionia*, with its only one species *W. saharae*, an endemic to the northwestern edge of the Sahara desert (Figure 3), may be found in several localities on dry shale in Algeria and Morocco. The aromatic specie was reported for the first time in the region of Beni Ounif (SW Algeria) by J. P.Adrien Warion (1837-1880), a French military physicianand botanist who made extensive collections while stationed in North Africa. The generic name of this plant is derivedfrom its last name. *W. saharae* grows on slopes of the Saharian Atlas (western Algeria) and in desert areas on basic and siliceous rocks from 3 to 500 m. This is a thistle-like aromatic plant, of 1 to 3 meter(s) in height, with white latex and fleshy, pinnately- partite leaves, the flowering season has been recorded from April to June [1, 9, 13].

- Kingdom :Plantae
- Division:Angiosperm
- Class:Eudicots
- Subclass : Asterids
- Order:Asterales
- Family:Asteraceae
- Subfamily:Cichorioideae
- Tribe:Cichorieae
- Sub-tribe:Warioniinae
- Genus:Warionia
- Specie : *Warionia saharae*
- Vernacular name: *Efessas, Kabarlemaiz, Abessas, Afezded* N° POSL Herbarium: CA02/07

Figure 3. Botanical classification of the medicinal plant *Warionia saharae* - Algeria Sahara.

According to classification system on flowering plants, the monospecific genus Warionia is closer to Cichorieae than to any other tribe of Asteraceae according to molecular and morphological characters, but it is so distinct from all other genera within the tribe that it requires a separate sub-tribe

Warioniinae. The genus and subtribe is characterized by a frutescent habit, latex, essential oils, the presence of both "oil ducts and latex" canals in the roots, homogamous capitula with slightly zygomorphic 5—dentate, tubular, yellow flowers with 10 corolla bundles, densely pilose achenes with a pap-pus of coarse, scabrid bristles, and a basic chromosome number of x =17 [20-22].

3. Ethnobotanical and Bioactivity of the Endemic Asteraceae Species

The ethnobotanical studies are recognized as the most viable method of identifying traditional drugs for bioactive compounds.Thus, the World Health Organization (WHO) estimates that 80% of people in developing countries depend on traditional medicine for their primary care. Different regions of Algerian Sahara are characterized by a high socio-cultural mixture and traditional therapeutics is mostly based on the contributions of several ethnics (Arab, Amazight, Cheleuh, Mozabit and Toureg). Therefore, this mixture of cultures offers a wealth of knowledge in uses of plants by the population in healthcare. The nomadic populations from Saharan who have livelihoods well-adapted to this hostile natural environment, and depend on plant as a source of food, medicine and fodder. So, the natural habitats of the Sahara are used for grazing, mainly by camels. Therefore, about five hundred species of higher plants inventoried part is nowadays used by people as medicinal plants in the Sahara [7, 10, 23, 24].

It is well known that species from Asteraceae family are used as natural remedies. Several Saharan plants, especialy from Asteraceae family have been and are still used as medicinal plants, particularly in folk medicine as stomachic, for treating diarrhea, gastrointestinal tracts, as antiinflammatory, antidiabetic, for skin diseases, insecticides and used as a food. Several studies on Saharn species have revealed very interesting biological activities such as antimicrobial, antiparasitic, antiviral, anti-inflammatory, antioxidant, antitumor capacities and molluscicidal activities especially of Asteraceae species which contained a diversity of chemical compounds such as diterpenoids, sesquiterpene lactones, triterpenoid saponin flavonoids, flavones glycosides, polyacetylenic fatty acids,and pyrrolizidine alkaloids [25].

In our previous ethnopharmacological works on Saharan traditional ethnopharmacopeae; We have noted that, the usually used plants from the south west Algeria region are from families of Apiaceae, Asteracea, Brassicaceae, Lamiaceae, Plumbaginaceae and Zygophelaceae. The

Medicinal plants from these families were used to treat more than one health problem and are often used for treatment of different illness such as uro-genital tract, gastrointestinal tract, respiratory tract, influenza infections, skin ailments, rheumatism, nervous system, cardiovascular system, inflammations and parasitic induced ailments [9, 26-28].

3.1. *Asteriscus graveolens*

Asteriscus graveolens is used in Sahara folk medicine as infusion or decoction for treating fever, gastrointestinal tractcomplaints, diarrhea, diabetes, cephalic pains, blennorrhagia, female infertility, infected wounds, sinusitis, bronchitis and as anti-inflammatory. The sap of the fresh leavesis used as drops for the nose and poultice for headaches [10, 9, 17].

The essential oil from this aromatic medicinal specie has been found to display a variety of biological activities that might be a natural potential source of bioactive compound used in food, cosmetics and in pharmaceutical products [29, 30-35]. Several biological properties have been repotted from extracts of *A. graveolens* such as: antimicrobial, antifungal and antioxidant activities. In year 2006, we have obtained an interesting antimicrobial activity of the essential oil from this specie against bacteria (*Escherichia coli, Staphylococcus aureus*, *Pseudomonas aeruginosa*, *Klebsiella Pneumaniae, Bacillus steorothermophilus* and *Enterococus Faecalis*) and fungus (*Candida albicans* and *Aspergillus niger*) [30]. The essential oil of *A. graveolens* shows a large zone of inhibition with *Pseudomonas aeruginosa* at the concentration of 25μg/ml.This oil also has a large inhibition zone against *Escherichia coli* at the same concentration. The essential oil is effective against gram-positive bacteria, with remarkable inhibitions zone for *S. aureus* and *B. steorothermophilus* [29].

The essential oil from the Moroccan specie was tested against three agricultural pathogenic fungi: *Penicillium digitatum, P. expansum* and *Botrytis cinerea*. The essential oil at 125, 150, 200, 250, 500, 1000 and 2000 ppm was highly effective against mycelial growth of *P. digitatum* with 100% inhibition from the first day of incubation. Complete inhibition was observed at 150, 200, 250, 500, 1000 and 2000 ppm concentrations of the essential oil from the first day of incubation for *P. expansum*. In addition, the essential oil was also tested at different concentrations on Citrus fruits (*Citrus reticulata* Blanco cv. Nules) inoculated with *P. digitatum* (105 conidia ml^{-1}, giving a fungistatic or fungicidal effect. In the same work, the team studie the antioxidant activity by using the DPPH test. The essential oil gave a value of IC_{50} = 0.2498 mg/ml, slightly lower than that of BHT (the control compound) for all concentrations used [32]. In addition, another study shows thatthe essential oil has an effective antifungal against *Alternaria sp, P. expansum,* and *R. stolonifer* and can be exploited as an ideal treatment against disease rot of apples or as a new potential source of natural additives for the food. [33].

An interesting biological activities were found for the essential oil (EO) obtained from *A. graveolens* harvested from Laghouat district (Algerian Sahara), where the authors of the study indicate that the EOs presented a weak scavenging activity (420.16 mg/mL), it exhibited good reducing power using the Phosphomolybdenum assay (0.28 AAEC/mg). The most important antibacterial activity was noted for *Bacillus cereus*. The oil revealed a remarkable activity against the tested fungi species with percentage inhibition up to 94.12% for *Fusarium culmorum* (BTCR). In their works, authors investigated for the first time the anticancer effect of the EO on two types of cancer cell lines (human liver carcinoma (HepG2) and Rat pheochromocytoma (PC12) cell lines). The EO showed anticancer activity against HepG2 cell at100μg/mL with inhibition of 82.98% of cells growth. The concentrations providing 50% of cell growth inhibition were less than 6.26 μg/mL for the PC12 cell line and more than 50 μg/mL for the HepG2 [36].

The organic extracts with different quantities (200, 400, 800 and 1,600μg) from aerial part and roots of*Asteriscus graveolens* have been used for controlling bayoud disease of date palm caused by *Fusarium oxysporum* f. sp. albedinis Foa. The antifungal test was conducted using disc diffusion technique and relative virulence (RV) test (on potato tuber tissue). The relative virulence was presented as necrotic tissue weight (mg) of potato tuber tissue. Among all solvents, methanol had the best extraction yield and the highest frequency of antifungal effect on Foa was presented by ethyl acetate extracts [37]. Recently, in our studie of antioxidant of Saharan flora by electrochemical methods, we have optimized the reducing power for antioxidant activity test on *A. graveolens,* by introduction the time interaction between the oxidant and the antioxidant in the reaction mixture as a new parameter. In this methodological approach the optimum time for our plant extract is 30 minutes [38].

Finally, Natural compounds isolated from *Asteriscus graveolens* presents a potential antitumor, antispasmodic, anti-inflammatory activity and are known also for the treatment of infected wounds. *A. graveolens* is also used as a natural active ingredient in some cosmetic products such as IBR-Pristinizer which is commercialized and used to protect cells from air pollution (IBR) [33, 34].

3.2. *Launaea arborescens*

Launaea arborescens (Batt) commonly used in Saharan popular medicine for treating diarrhea, gastrointestinal tracts, hepatic pains, children fever, as anti-inflammatory, for skin diseases, as soporific, lactagogue, diuretic, anti-tumour, for treatment of infected wounds, and as insecticidal. The latex is applied locally to cure sore throats and in the treatment of furuncles. The powdered root mixed with Artemisia herba-alba is taken for diabetes. Honey from bees that have been on the L. arborescens flowers is considered fortifying. The plant is appreciated by livestock, mainly by camel [2, 7, 10, 30].

The methanol extract of the aerial part of *Launaea Arborescens* showed very interesting antifungal activity against Candida albicans and Saccharomyces cerevisae and antibacterial activity against *Staphylococcus aureus, Escherichia coli, Pseudomonas aeruginosa* and *Klebsiella entrecocus* [39, 40]. Aerial part and roots of *Launeae arborescens* were used to evaluate their extracts for antifungal activity against Fusarium oxysporum f. sp. albedinis Foa. [37]. In the agricultural field, the Saharan termite, Anacanthotermes ochraceus is among the most devastating termite pests, it cause great damage to house hold materials and agriculture in oasis. In our research on bioinsecticides from Saharan plants, among plants tested as termiticidal and eco-friendly termite control agents, aqueous extracts from aerial part of *Launaea arborescens* showed very moderate anti-termitic activity against termite workers in a different doses (lethal effect appears after 4h with 50% mortality for concentrations of 2%, 3%, 4% and 5%,), the observed activity is less than found in tested toxic saharan plants (*Calotropis procera, Hyoscyamus muticus, Pergularia tomentosa* and *Datura stramonium*) [41, 42]. On another hand, in terme of toxicity, the hydroalcoholic extract from aerial parts of Launaea arborescens was evaluated for acute and subacute toxicity in Swiss mice after ingestions of the extract. The LD_{50} of the extract is higher than 2.75 g/kg and the subacutetreatment did not shows any change in corporal weight and haematological parameters, suggest that the plant seems to be destituted of toxic effects in mice [43].

3.3. *Warionia saharae*

The specie *Warionia saharae* is considered to have very interesting place in the saharan traditional ethnopharmacopeae. Decoction of dried leaves is used as antirrheumatic, anti-inflammatoryand against gastrointestinal tracts, icter and epileptic crisis. In addition, the local women anoint themselves with the perfume of the leaves, and believe that the supernatural powers attributed to the plant make them more seductive [7, 9, 44].

Almost ten years ago, we have investigated the antibacterial and antioxidant activities of *W. saharae* harvested from Algerian Sahara. So, liquid-liquid fractionalization of the crude 80% EtOH extract along with $CHCl_3$, EtOAc and *n*-BuOH extracts from leaves of *W. saharae* were tested for antioxidant activity by usingDPPH and superoxide anion radical. We found that the AcOEt extract has an important role in scavenging abilities of various radicals and IC50 values of antioxidant activities followed 3.08 ± 0.40 µg/mL for DPPH and 8.95 ± 0.23 µg/mL for superoxide anion radical. The oil-ethanol solution in three dilutions (1/2, 1/4 and 1/8 v/v, in the absolute ethanol), was tested against five bacteria (*Staphylococcus aureus, Escherichia coli, Klebsiella pneumaniae, Pseudomonas aeruginosa* and *Enterococcus faecalis*), and two fungus (*Saccharomyces cerevisiae* and *Candida albicans*) [45, 46].

A broad spectrum of antibacterial activity was observed, thus, the essential oils (1 mg/mL) from Moroccan specie exhibited a strong antibacterial effect as a diameter of zones of inhibition (28,5 ± 2,12 and 37,5 ± 3.53 mm) against *St. aureus* and *P.aeruginosa*, respectively, while, ethanolic extract (50 mg/ml) exhibited a moderate effect against tested bacterial. MICs values of oil and the extract were ranged 0,031-0,25µg/mL and 6,25-12,5 mg/mL, respectively. The same team demonstrate by using in vitro bioanalytical methods (DPPH, PPM and FRAP) for the evaluation of the antioxidant and free radical scavenging abilities, that the IC_{50} value of *W. saharae* of saponins extract was weak (IC_{50} = 74.14 µg/mL) with an antioxidant activity higher than the ethanolic extract (IC_{50} = 182 µg/mL) and ethyl acetate extract (IC_{50} = 197.4 µg/mL) [47]. Essential oil from the aerial parts of *W. saharae* inhibited significantly the mycelial growth of phytopathogenic causing the deterioration for apple ($p<0.05$). Fungal spore production was completely inhibited at 1 µL/mL air for *Alternaria sp.* and at 2 µL/mL air for *Penicillium expansum* and *Rhizopus stolonifer* [48].

Ethanolic extract from *W. saharae* at oral dose of 5 g/kg BW inducted nephrotoxicity and caused the damaged structural integrity of the liver. The ethanolic extract at doses of 200 and 400 mg/kg BW decreased the inflammation (56.42% and 61.45% inhibition, respectively) from the 2ed hour after carrageenan administration. Ethanolic extract has shown significant ($p < 0.05$) inhibition of paw edema when compared with control rats (NaCl 0.9%) and were higher to than the reference drug (Diclofenac) [49].

Finally in term of biological activities, we noted that our three endemic species exhibited an interest in anti nosocomial and molluscidal activities. Extracts from *Asteriscus graveolens*, *Launaea arborescens* and *Warionia saharae* shown an antimicrobial effect against two nosocomial strains of the genus *Staphylococcus* (*Staphylococcus aureus* and *Staphylococcus epidermidis*). We foundan interesting anti nosocomial activity for *Asteriscus graveolens* with maximum inhibition zones of 22 mm for S. aureus but only 12 mm for *Staphylococcus epidermidis*. *Launaea arboresens* showed the smaller maximum inhibition zones. *Warionia saharae* showed the minimal inhibitory concentration of 7.10^{-4} g, followed by *Launaea arboresens* (9.10^{-4} g) and *Asteriscus graveolens* with the concentration of 10^{-3} g. We noted that the *Staphylococcus* strains tested were resistant to tetracyclines, sulfamides and β lactams with successive rates of 100%, 67.70% and 51.84% whereas their rates of resistance for tested plants were 30.92% for *Launaea arboreseens* [50].

In another studies, we have found that extracts of the aerial parts of *Asteriscus graveolens, Launaea arborescens and Warionia saharae* showed significant molluscicidal against *Lymnaea acuminata* with a median lethal concentration (LC_{50}) of aqueous extract (8.178 μg/mL) and organic extract 0.002 μg/mL, which was indicated higher potency than the positive control, (LC_{50}=100 μg/mL for aqueous extract; LC_{50}=11.6μg/mL for organic extract). Particularly, the dichloromethane soluble fraction showed the most potent molluscicidal activity against *Lymnaea acuminate* and the acetone extract of the aerial parts of *Warionia saharae* exhibit significant molluscicidal activities [51].

4. Terpenoids of the Endemic Asteraceae Species and Their Pharmacological Activities

The isolation and structural identification of bioactive natural compounds is one of the most important areas of research in phytochemistry. Furthermore, only a small part of the 400.000 vegetable species known were investigated from phytochemical and pharmacological aspects. In another hand, the determination of molecular structure of the bioactive natural compounds, have an important impact on the knowledge in the field of chemistry, biology, pharmacology by the production of synthetic synthon, incorporation of pharmacomodulation and rationalization of mechanism of action.Technically, the obtaining of pure bioactives natural natural compounds can be achieved by chromatographic techniques (CC; TLC; GC; CCC; CE; HPTLC; HPLC; HPCPC; FCPC….) as effective methods of separation and purification, usually linked with an online spectroscopic (GC-MS; GC-MS-MS; LC-MS; LC-MS-MS; SPE-LC-MS; LC-PDA-MS; LC-HMRS;HPLC-NMR; LC-NMR-MS…) [1, 2].

More than 100,000 terpenoids (almost 1/3 of the exiting secondary metabolites) have been identified from the various parts of medicinal plants with particular chemical structures and diversity of skeletal as isoprene-based compounds, forming monoterpenes (C10), sesquiterpenes (C15), diterpenes (C20), sesterterpenes (C25), triterpenes (C30), and polyterpenes (>C30). The biosynthesis of terpenoids follows two major pathways: Mevalonic acid (MVA) pathway and 2C-methyl-D-erythritol-4-phosphate (MEP) pathway. In plants; terpenoids are constituents of resins, waxes, rubber and balsams, while essential oils are mainly composed of mono- and sesquiterpenes. These natural compounds are constituents of the cell membrane of plants, and give plant odor specificity and they play important roles in cell membrane defense acting as allelochemicals to attract pollinators, repel or attract herbivore predators [2, 52].

Considerable attention has been given to Saharan medicinal plants that contains many chemical compounds with therapeutic benefits such as terpenoids isolated from extracts and essential oils. In the follows we indicate the different classes of terpenoids isolated and identified in the three endemic medicinal plants: *Asteriscus graveolens*, *Launaea arborescens* and *Warionia saharae*.

4.1. Triterpenes

Triterpenes represent a special and interesting class of natural compounds, particularly when they form the saponins derivatives. A great interest was done to their pharmacological properties, role in the defense and adaptation system of the plant and and their primary role in the plant metabolism like the phytosterols. More than 20,000 triterpenoids in various subclasses have been reported in the literature. Biochemically, triterpenes arise from dimerization of two farnesyl pyrophosphate units that produce the intermediate compound squalene. The majority of the triterpenoides are pentacyclic and belong to lupane, gammacerane, oleanane, and ursane groups, with some tetracyclic compounds such as α-amyrin **1**, β-amyrin **2,** ψ-taraxasteryl **3**, taraxerol **4**, taraxasterol **5**, lupeol **6.**

Biologically, it is demonstrated that taraxerol **4** andtaraxasterol **5**exhibited potent anti-tumour promoting activity in carcinogenesis tests of mouse skin. In addition they had an inhibitory effect on mouse spontaneous mammary tumours, indicating that terpenoids with anti-tumour effects can be potentially anti-carcinogenic [53].

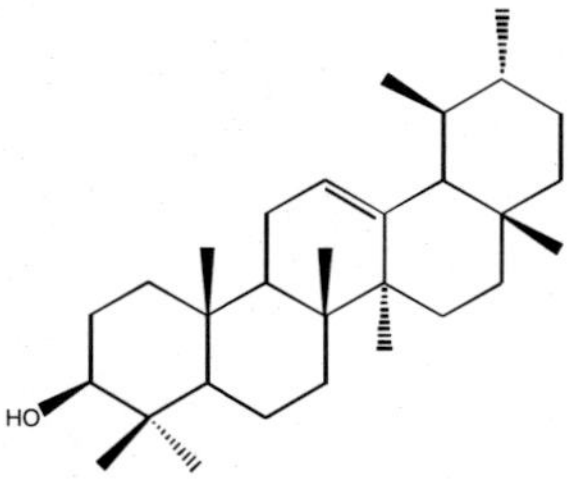

1

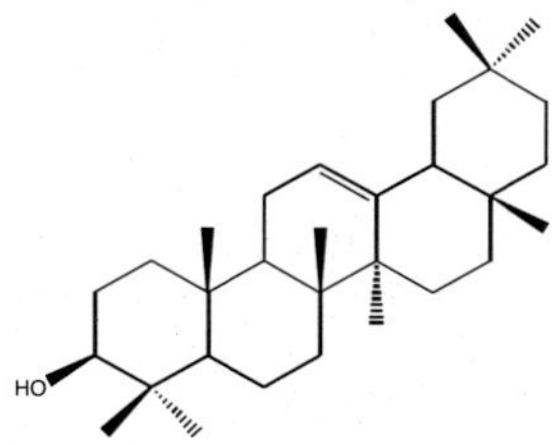

2

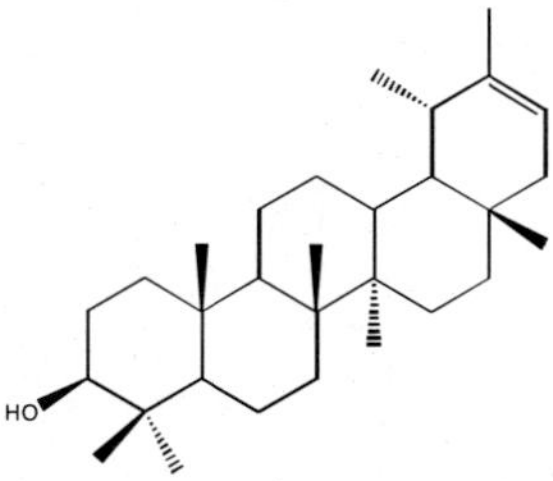

3

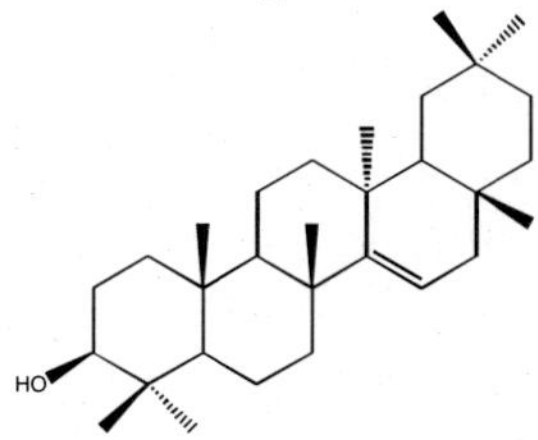

4

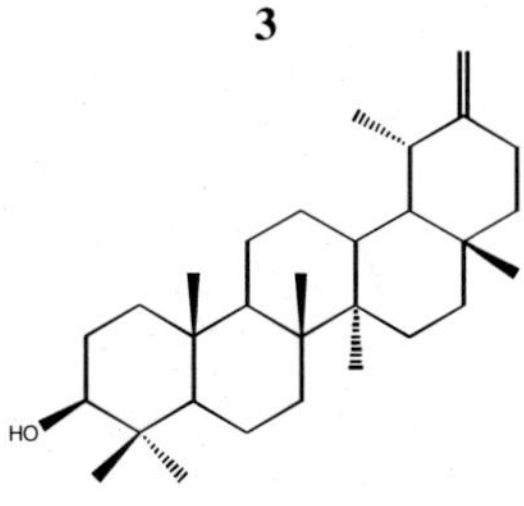

5

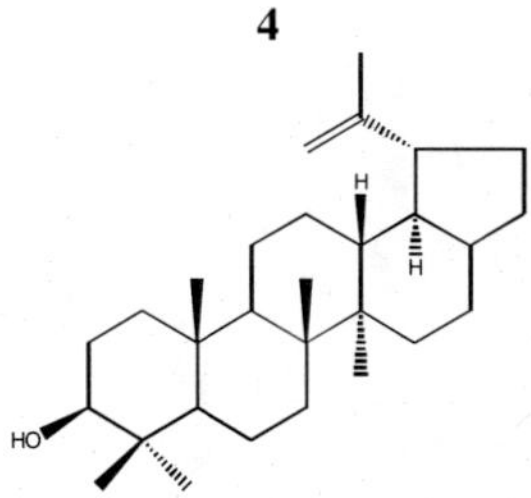

6

Figure 4. Structure of some important natural triterpenes.

A diversity structure of triterpenes oleanane, sesquiterpenes are the main terpenoids isolated and chemically characterised from *Asteriscus graveolens*,

Launaea arborescens and *Warionia saharae.* Triterpenes are identified particularly in *L. arborescens.* Moreover, the works carried out on the other two species *A. graveolens* and *W. saharae*, mainly shows the presence of sesquiterpenes.

The first phytochemical study on *Launaea arborescens* collected in Algerian Sahara dates back 15 years, exactly in 2006, we have described the isolation from the methanol extract of the aerial parts of this specie, four compounds: the first diterpene, methyl-15,16-epoxy-12-oxo-8(17),13(16),14-ent-labdatrien-19-oate **7** isolated from this Saharan specie, with two flavonoids and one lignin derivative [54].

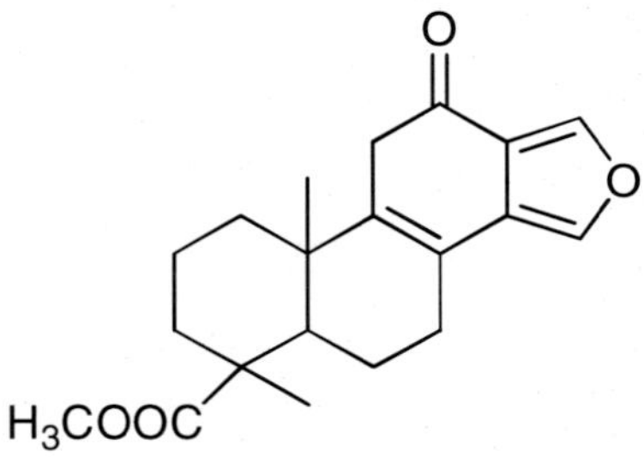

7

Figure 5. Structure of the diterpene identified in *Launaea arborescens.*

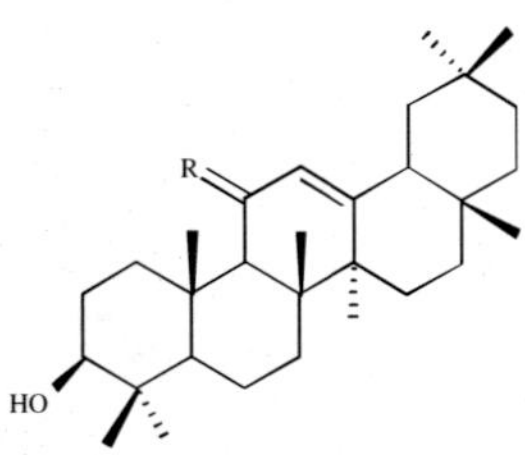

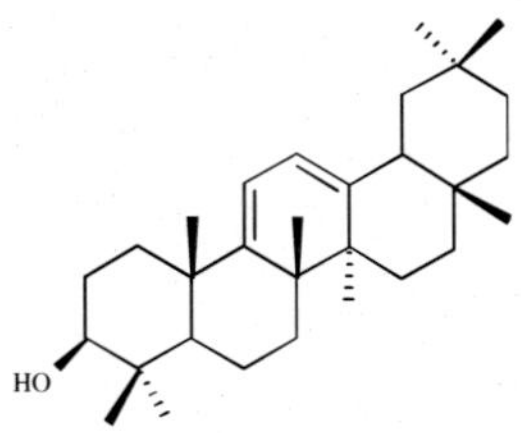

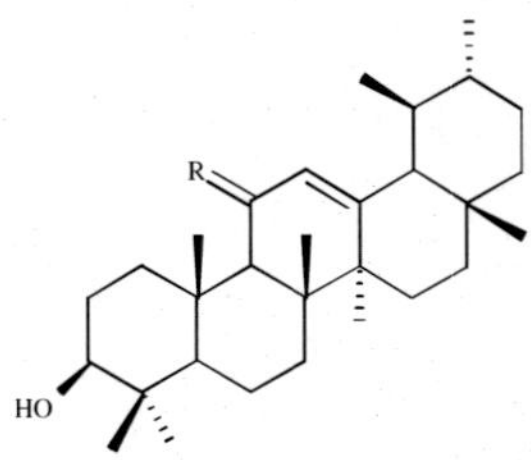

8-11 **12** **13-16**

8 R= α OC_2H_5, β H
9 R= α OCH_3, β H
10 R= α OH, β H
11 R= O

13 R= α OC_2H_5, β H
14 R= α OCH_3, β H
15 R= α OH, β H
16 R= O

Figure 6. Structures of triterpenes identified in *Launaea arborescens.*

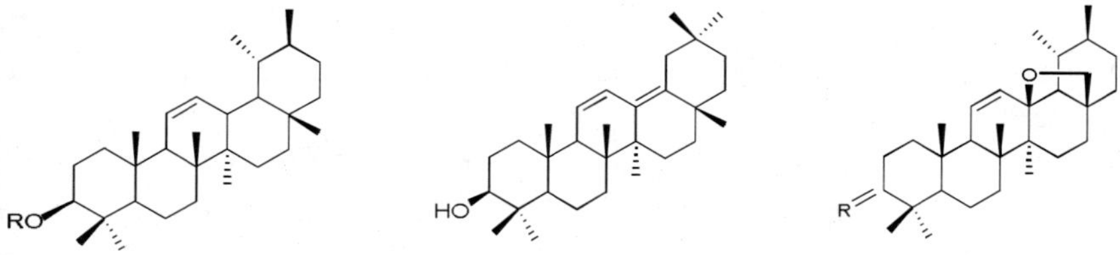

17-18 **19** **20-21**

17 R= OH
18 R = Ac

20 R= α H, β OH
21 R = O

Figure 7. Structures of triterpenes identified in *L. residifolia* and *L. nudicaulis.*

Two years after, in 2008, Bitam et al. [55] while studying with Italian team the Algerian specie; have isolated from both the aerial parts and roots of *L. Arborescens* a series of triterpenes oleanane such as: 3β-hydroxy-11α-ethoxy-olean-12-ene **8**, 3β-hydroxy-11α-methoxyolean-12-ene **9**, 3β-11α-dihydroxy-olean-12-ene **10**, 3-β-hydroxy-11-oxo-olean-12-ene **11**, oleana-9(11):12-dien-3β-ol **12**, 3β-hydroxy-11α-ethoxyurs-12-ene **13**, 3β-hydroxy-11α-methoxyurs-12-ene **14**, 3β-11α-dihydroxy-urs-12-ene **15** and 3-β-hydroxy-11-oxo-urs-12-ene **16**.

As an indication, different triterpenes are isolated from other medicinal *Launaea.* Thus, from *L. residifolia* (L.) four triterpenes are isolated: α-amyrin **1**, lupeol **6**, lupeol acetate and their epimer moretenol. Chemical investigation of *L. nudicaulis* yielded also α- amyrin **1**, lupeol **6**, together with triterpenes such as: ψ- taraxasterol **3,** 3β- taraxerol **4**, taraxasterol **5**, nudicauline A **17**, and nudicauline B **18**, olean-11,13(18)-diene **19**, 3β-hydroxy-13(28)-epoxy-urs-11-ene **20** and 3-keto-13(28)-epoxy-urs-11-ene **21** [2, 56]

4.2. Sesquiterpenes

Sesquiterpenoids resulting from the combination of three branched, unsaturated isoprene units (C5) represent a large class of natural products with a wide range of biological activities. Thus, more than 4000 sesquiterpenoids structures with around 30 different skeletal types have so far been reported from species of Asteraceae family.

These natural compounds exhibit a wide range of bioactivities which include plant growth regulation, antimicrobial, antiparasitic, anthelmintic, antiphlogistic, antispasmodic, as schistosomicidal and insect feeding-deterrent agents. Also they provoke the toxicity for certain cancer cell lines, and the particularity of sesquiterpene lactone type parthenolide are used in the treatment of fever, migraine headaches, rheumatoid arthritis, as an anti-inflammatory agent and shown potent anticancer activity [2, 57].

Among the specific characteristics of the sesquiterpene lactones class, is their use for chemo-systematic studies within the botanical family. Due to their interesting chemical diversity as chemio-characteristic of Asteraceae family, below is listed some typical members of sesquiterpene lactones [2, 39] (Figure 8).

- Guaianolides: Ixerisoside B **22**, Scorzoside **23**, Zaluzanin C **24**, and Prenantheside C **25**.
- Guaianolides: Lactucin **26**, Crepidiaside A **27**, 8-O-acetate,11β,13dihydrolactucin **28** and 8-Deoxylactucin **29**.
- Eudesmane: Santamarin **30**, Ixerisoside E **31**, Sonchuside C **32** and Artesin **33**
- Germacranolides: Picriside B **34**, and Sonchuside A **35**
- Melampolides: Lactulide A **36** and Lactuside A **37**

The occurrence of terpene-rich latex is a one of the particularity of Cichorieae tribe (Asteraceae family) including *Launeae arborescens* and *Warionia saharae.* Thus, the ethyl acetate extract of the roots of *L. arborescens* is rich on antimicrobial sesquiterpene derivatives exhibiting eudesmane, guaiane and germacrane skeletons, and identified as 9α-hydroxy-11β,13-dihydro-3-epi-zaluzanin C **38**, 9α-hydroxy-4α,15-dihydrozaluzanin C **39**, 3β,14-dihydroxycostunolide -3-O-β-glycopyranoside **40**, 3β, 14-dihydroxy costunolide -3-O- β-glucopyranosyl-14-O-p-hydroxy phenylacetate **41** and a sesquiterpenoid sulphate, 8-deoxy-15- (3'-hydroxy-2'-methylpropanoyl) – lactucin -3'sulfate **42** [55, 58] (Figure 9).

Much interesting work on the phytochemistry of extracts obtained from leaves of *Warionia saharae*. So, diversity structures of sesquiterpene lactones are isolated from the MeOH-soluble fractionation part of the dichloromethane extract obtained from *W. saharae* leaves. Six oxygenated guaianolides **43-48** were identified and exhibiting a significant cytotoxicity against the KB cancer cell line, with IC_{50} values respectively 1.0, 4.5, 1.7, 2.0, 3.3, and 5.5 mg/mL.

It has been showed that no significant differences in cytotoxicity were observed between two compounds varying only in the position of the double bond (9,10 or 10,14). In contrast, acylation of one of the free hydroxy groups resulted in a clear increase of activity. Also, two other cytotoxic dimeric sesquiterpene lactones **49** and **50** were isolated [59-61].

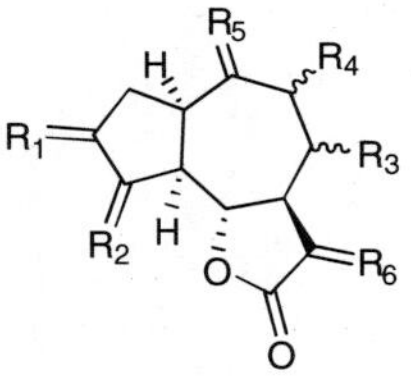

22-25

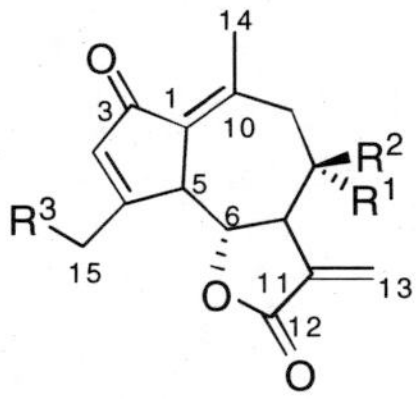

26-27

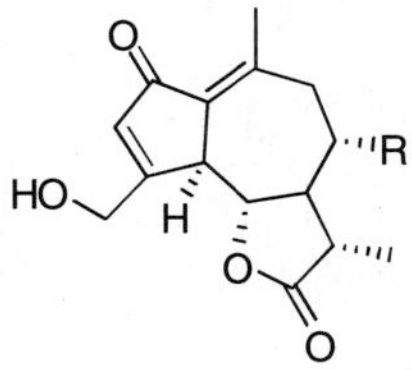

28-29

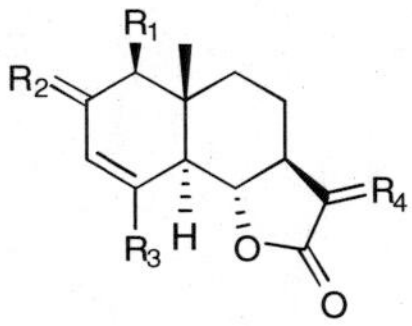

30-31

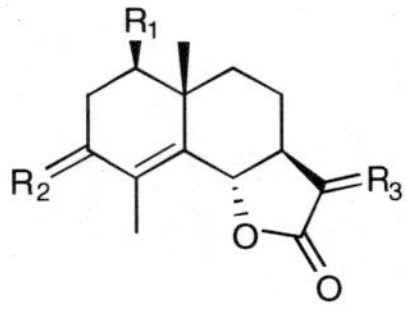

32-33

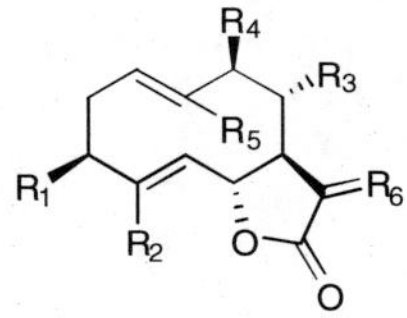

34-35

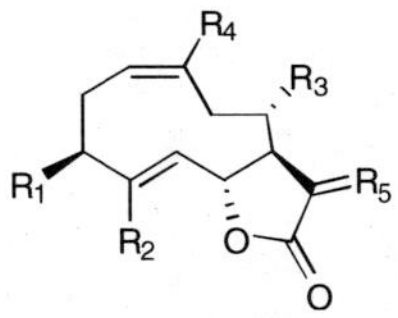

36-37

	R_1	R_2	R_3	R_4	R_5	R_6
22	α H, β OGlc	α CH_3, βH	β OH	H	CH_2	CH_2
23	H,H	CH_2	H	α OGlc	CH_2	α CH_3, βH
24	α H,β OH	CH_2	H	H	CH_2	CH_2
25	α H,β OGlc	CH_2	α OH	H	CH_2	α CH_3, βH
26	OH	H	OH			
27	H	H	O-glc			
28	OAc					
29	H					
30	OH	H, H	CH_3	CH_2		
31	OH	H, H	CH_2OGlc	CH_2		
32	H	α H, β OGlc	α CH_3, β H			
33	OH	H, H	α CH_3, βH			
34	H	CH_2OGlc	H	H	CH_3	CH_2
35	O-Glc	CH_3	H	H	CH_3	α CH_3, βH
36	OH	CH_3	H	CHO	α CH_3, βH	
37	OGlc	CH_3	H	CHO	α CH_3, βH	

Figure 8. Structures of typical members of sesquiterpene lactones identified in Asteraceae family.

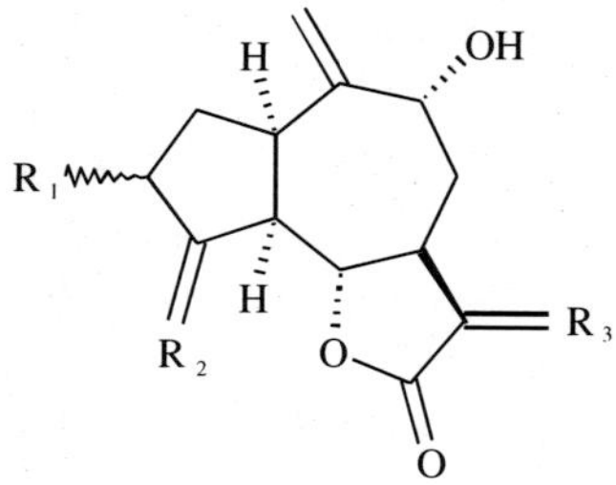

38-39

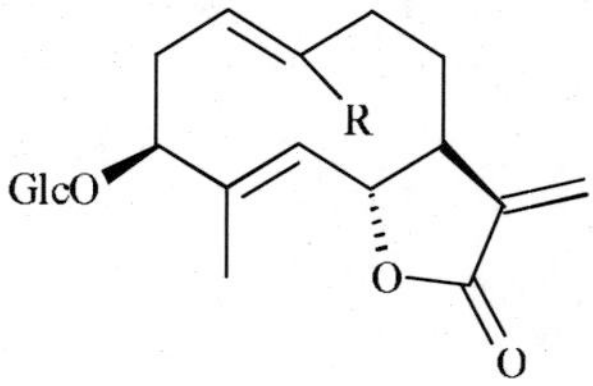

40-41

	R_1	R_2	R_3
38	α OH	CH_2	α CH_3, β H
39	β OH	β CH_3	CH_2

	R
40	CH_2OH
41	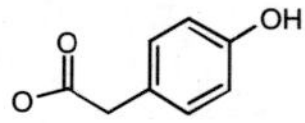

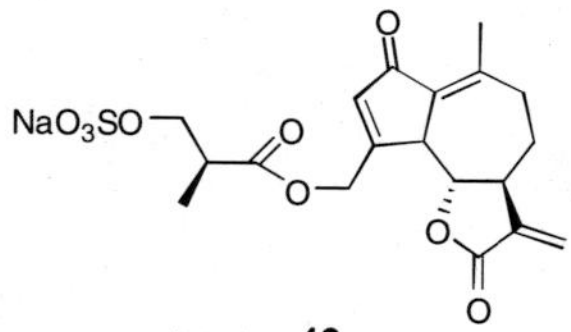

42

Figure 9. Structures of sesquiterpene lactones identified in *Launaea arborescens* roots.

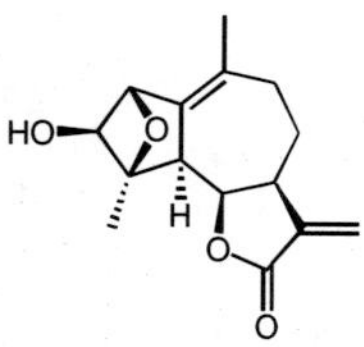

43

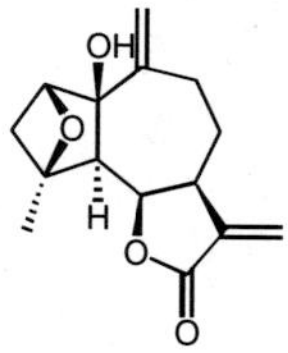

44

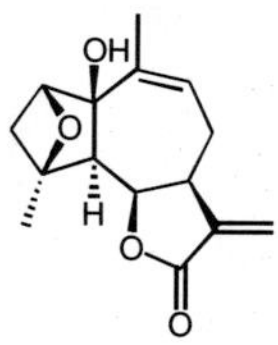

45

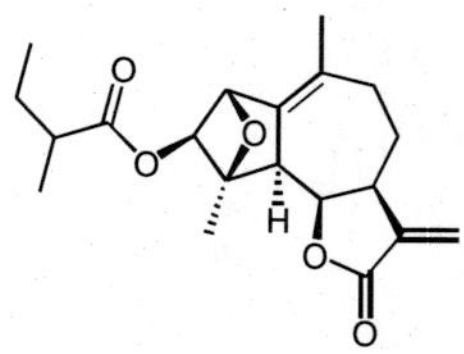

46

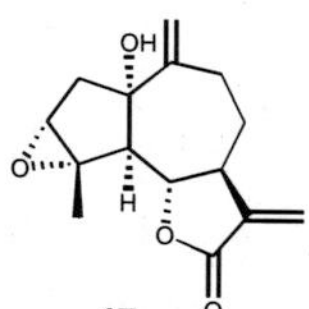

47

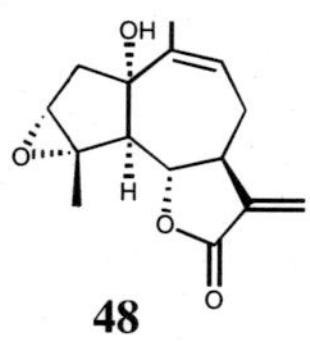

48

Figure 10. Structures of oxygenated guaianolides identified in *Warionea saharae.*

The same team [61] has explored the methanol soluble part of the dichloromethane extract of the leaves of *Warionia saharae*, and has isolated a serie of guaianolide-type sesquiterpene lactones **51-55** and eudesmane type sesquiterpene **56.**

The isolated guaianolide type sesquiterpene lactones from *W. saharae* were examined for their cytotoxic and anti-inflammatory effects in HeLa, Jurkat T and human peripheral blood mononuclear cells. Guainolide lactones tested were found to exert a strong cytotoxicity similar in potency in all investigated cell types. In addition, somes guainolide lactones showed a potent and comparable down-regulation of the mRNAs of the house-keeping genes beta-actin and GAP-DH in PBMCs after 20 h. Bioactive guaianolides were also subjected to an IL-6 luciferase reporter gene assay and showed IC50 values of 1.0 and 1.2 microM [59, 60].

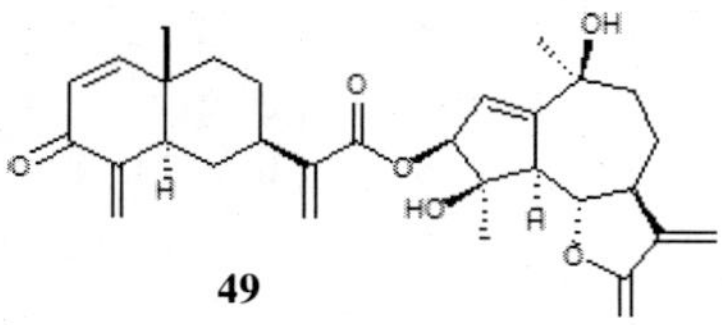

49

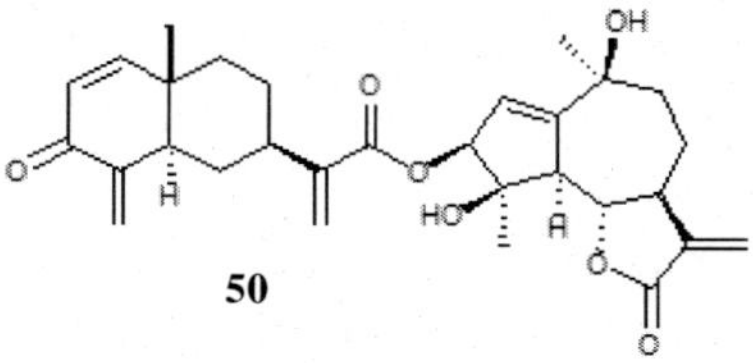

50

Figure 11. Structures of dimeric sesquiterpene lactones identified in *Warionea saharae.*

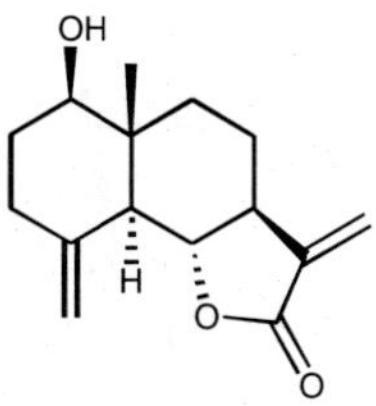

51

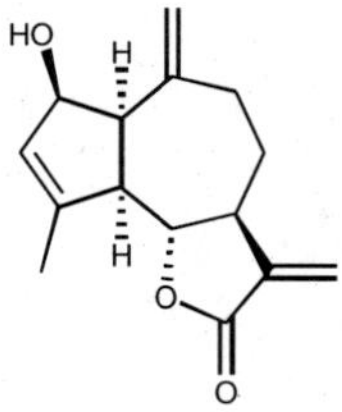

52

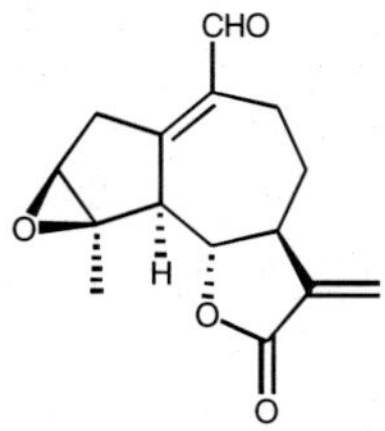

53

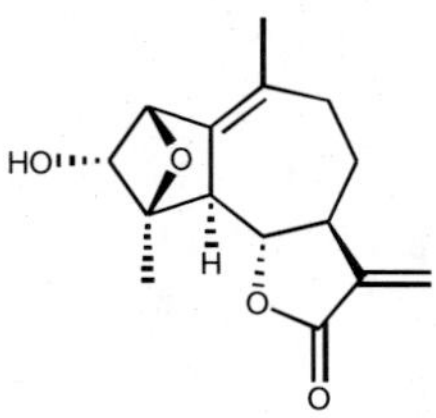

54

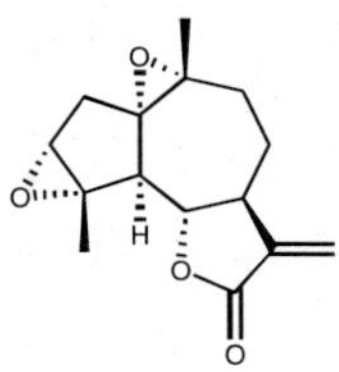

55

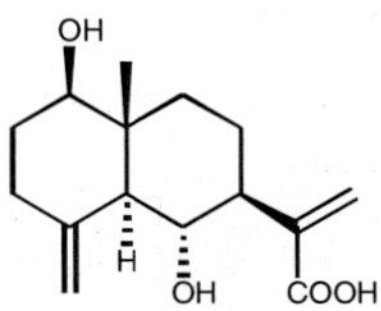

56

Figure 12. Structures of guaianolide and eudesmane type sesquiterpene identified in *Warionea saharae.*

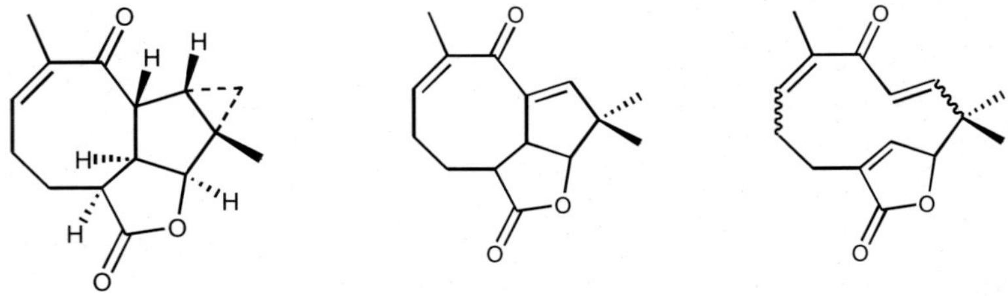

57 **58**

	Δ^6	Δ^9		Δ^6	Δ^9
59a	*Z*	*Z*	**59c**	*Z*	*E*
59b	*E*	*Z*	**59d**	*E*	*E*

Figure 13. Structures of sesquiterpene lactones identified in *Asteriscus graveolens.*

The occurrence of sesquiterpenoids is also reported in the aerial parts of *Asteriscus graveolens*, where five sesquiterpene lactones have been identified as: naupliolide **57,** 6,7,9,10-tetradehydro-asteriscanolide **58** and asteriscunolides **A–D (59a–d)** [62].

Finally, the current COVID-19 pandemic which caused over 220 million infections resulting in over 4 millions, since it was declared a pandemic by World Health Organization (WHO) on March 11, 2020. WHO, has demanded quick research of new drugs, vaccines and potential therapeutic strategies. The structural information of the recently published crystal structure of the main protease COVID-19 (Mpro) in complex with an inhibitor has made it possible to launch docking works around the world on different natural or synthetic molecules. So, on our side we have utilized to conduct extensive virtual docking screening of a six cytotoxic oxygenated guaianolides **43**, **44**, **45**, **46**, **47**, and **48** identified in *Warionea saharae*. The three-dimensional (3 D) structures of compounds were fully optimized by using the DFT method with the B3LYP functional, and a standard valence double-zeta basis set, 6-311G**, was used for all kinds of atoms. Molecular docking analysis was used to study the type of interactions between all compounds (6 molecules) and the target (Coronavirus (2019-nCoV) spike protein). Interestingly, in our screening, the estimated free energy of binding varied in the range of -3.07 to -5.88 kcal/mol, suggesting, that molecules can spontaneously interact within the binding site of SARS-Cov-2 Mpro. The inhibition activity of oxygenated sesquiterpene guaianolide **44** was found to be the highest as its value (-5.88 kcal/mol) [63].

4.3. Volatile Terpenoids from Essential Oils

The three Saharan species *Asteriscus graveolens*, *Launaea arborescens,* and *Warionea saharae* are aromatic plants with a characteristic odors, very appreciated and having a privileged place in the local ethnopharmacopeae. On the scientific field, a great attention was paid for their essential oils [1, 9].

It is well known that essential oils from plants consist of chemical mixtures involving a several different kind of volatile molecules, mainly formed by terpenes (monoterpenes, sesquiterpenes, diterpenes hydrocarbons and their oxygenated derivatives) and phenylpropanoids (benzene derivatives) which are responsible for the characteristic flavor, aroma and the biological activities of the plants that contain.

In many spices, monoterpenesand their derivatives are reported to be the main constituents of essential oils extracted from herbs, spices, fruits, and vegetables. Monoterpenes as simple terpenoids contain 10 carbon atoms arranged in variouscyclic or acyclic branched isoprene units.These natural volatile compounds are an important class of commercially interesting terpenoids widely used as flavoring agents in foods and cosmetics, topical medicines, perfumes, insecticides and are known to exhibit a large range of biological activities [64].

Diversity of terpenoids with oxygenated monoterpenes as the main compounds are identified in the essential oils extracted from the three Saharan Asteraceae (*A. graveolens*, *L. arborescens, W. saharae)* pursched from different regions as shown in Table 1 and Figure 15.

In our work dating from 2007, we have indicated that the essential oils of the aerial part of the *Asteriscus graveolens* contained mainly oxygenated monoterpenes (37.6%, 60.3%) in leaves and flowers, respectively, with 1,8-cineole (21.5%) as the main constituent in the leaves and 2,6-dimethyl-1,6-heptadien-4-yl acetate (19.4%), *trans*-chrysanthenyl acetate (18.7%) and 1,8-cineole (16.5%) as the main constituents in the flowers. Oxygenated sesquiterpenes were also the dominant class of compounds in both leaves and flowers (27.6% and 15.2%, respectively), with δ-cadinol (19.1%, 13.9%) as the main constituent plant [17]. Following, the presence of chiral molecules in the essential oils from *A. graveolens*, we carried out a study on the chiral signatures by using Chiral HPLC with polarimetric detection and vibrational circular dichroism (VCD).

Table 1. Main Volatile terpenoids in essential oils of endemic Saharan Asteraceae

Species	Origin	Part of plant	Main compounds in EOs	Ref
Asteriscus graveolens	Algeria	Leaves	1,8-Cineole, δ-Cadinol	[17]
		Flowers	2,6-dimethyl-1,6-heptadien-4-yl acetate, *trans*-Chrysanthenyl acetate, 1,8-Cineole, δ-Cadinol	
		Aerial parts	*cis*-Chrysanthenyl acetate, *cis*-8 acetoxychrysanthenyl acetate, τ-Muurolol	[65]
			cis-Chrysanthenyl acetate, Myrtenyl acetate, Kessane, Liguloxide, δ-Cadinol, *cis*-Chrysanthenol, Valerianol	[66]
	Egypt	Aerial parts	1,5-dihydroxy-6,7-dimethylocta-3,5-diene, Carveol, Cedrenol, *α*-Pinene, *α*-Himachalene; *α*-Phellandrene	[67]
	Morocco	flowers	*cis*-8-acetoxychrysanthenyl acetate, *cis*-Chrysanthenyl acetate, Cadinol, 6- oxocyclonerolidol, *α*-oxobisabolene	[68]
		Aerial parts	6- oxocyclonerolidol, 6-hydroxycyclonerolidol, Intermedeol	[69]
			6-oxocyclonerolidol, Cadinol, *α*-Pinene, *α*-Bisaboloneoxide, HumuleneepoxydeII, Bisabolone	[70]
Launaea arborescens	Algeria	Aerial parts	1,8 Cineol, Sulcatone	[71]
Warionea saharae	Algeria	Leaves	Carophylane, Guaiene, Cadinene	[72]
			Eudesmol, Isomenthol, Terpinyl butyrate, Trans-nerolidol, Terpinen-4-ol, Linalool, Caryophyllane	[73]
	Morocco	Leaves	Eudesmol,Guaiol, Linalool, Nerolidol, Terpinen-4-ol, 1,8-Cineol	[74]
			Eudesmol, Linalool, Nerolidol	[75]
			Eudesmol, Nerolidol, Linalool, Camphor, Terpineol	[76]
			Eudesmol, Nerolidol	[77]
			Eudesmol, Nerolidol,Linalool, 1,8-Cineole	[78]

Three major chiral compounds were isolated by preparative HPLC. The naturally occurring oxocycloneroledol **60**, mostly found in the leaf oil (49.4-

55.6%), presents a (+) sign, but cis–chrysanthenylacetate **61** and cis–acetyloxychrysanthenyl acetate **62**, mostly found in the flower EO (35.9-74.9% and 10.0-34.3%, respectively), both present a (−) sign. The comparison of the experimental and calculated VCD spectra of pure isolated **60**, **61**, and **62** provided their absolute configuration as being [79].

60: (2R,6R)-(+) -6-ethenyl-2,6-dimethyl-2-(4-methylpent-3-en-1-yl) dihydro-2H-pyran-3(4H)-one)

61: (1S,5R,6S)-(−)-2,7,7-trimethylbicyclo[3.1.1] hept-2-en-6-yl acetate

62: (1S,5R,6R,7S)-(−)-7-(acetyloxy)-2,6-dimethylbicyclo [3.1.1] hept-2-en-6-yl] methyl acetate 3.

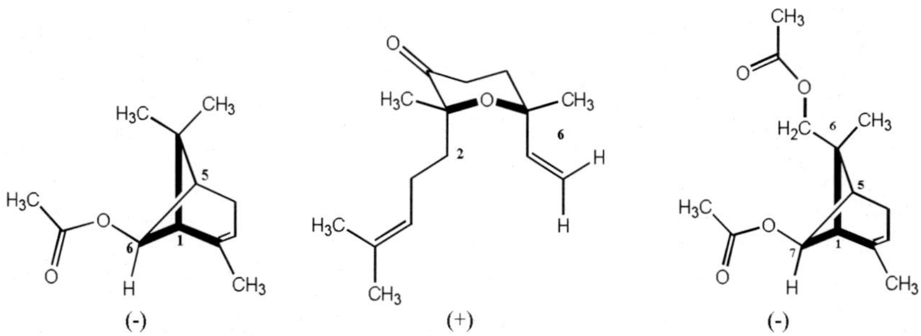

61 **60** **62**

Figure 14. Structures of the major chiral compounds from EOs of *Asteriscus graveolens.*

The GC/MS analysis of essential oils (EOs) from *A. graveolens* harvested from different eco-system, indicate a remarkable difference in the chemical composition of EOs. Works of Moroccan teams indicate a variability in the composition of EOs, thus, 6- oxocyclonerolidol (66.7%), 6-hydroxycyclonerolidol (8.8%) and intermedeol (4.3%) in *A. graveolens* aerial parts [69]; *cis*-8-acetoxychrysanthenyl acetate (48.5%), *cis*-chrysanthenyl acetate (13.4%), cadinol (11.4%), 6- oxocyclonerolidol (7.8%), *α*-oxobisabolene (5.7%) in *A. graveolens* flowers [68], 6-oxocyclonerolidol (30.72%), cadinol (14.50 %), *α*-pinene (4.22%), *α*-bisabolone oxide (3.56%), humulene epoxyde II (3.5%) and bisabolone (3.5%) in *B. graveolens* subsp. *odorus* inaerial parts [70]. An abundance of 1,5-dihydroxy-6,7-dimethylocta-3,5-diene (8.62%), carveol (6.07%), *α*-pinene (3.83%), cedrenol (3.10%), *α*-phellandrene (2.12%), and *α*-himachalene (1.89%) is observed in the Egyptian specie [67]. The major components in essential oilsfrom *Asteriscus graveolens* aerial parts harvested in Laghouat (Algeria) are *cis*-chrysanthenyl acetate (44.30%), *cis*-8 acetoxychrysanthenyl acetate (33.70%) and τ-muurolol (6.51%) [65]. Also, the specie from the extreme Algerian Sahara (Hoggar) is characteristic by an EOs rich on *cis*-chrysanthenyl acetate (31.1%), myrtenyl acetate (15.1%), kessane (11.5%), liguloxide (7.4%), *cis*-chrysanthenol (5.3%), δ-cadinol (4.5%) and valerianol (3.8%) [66].

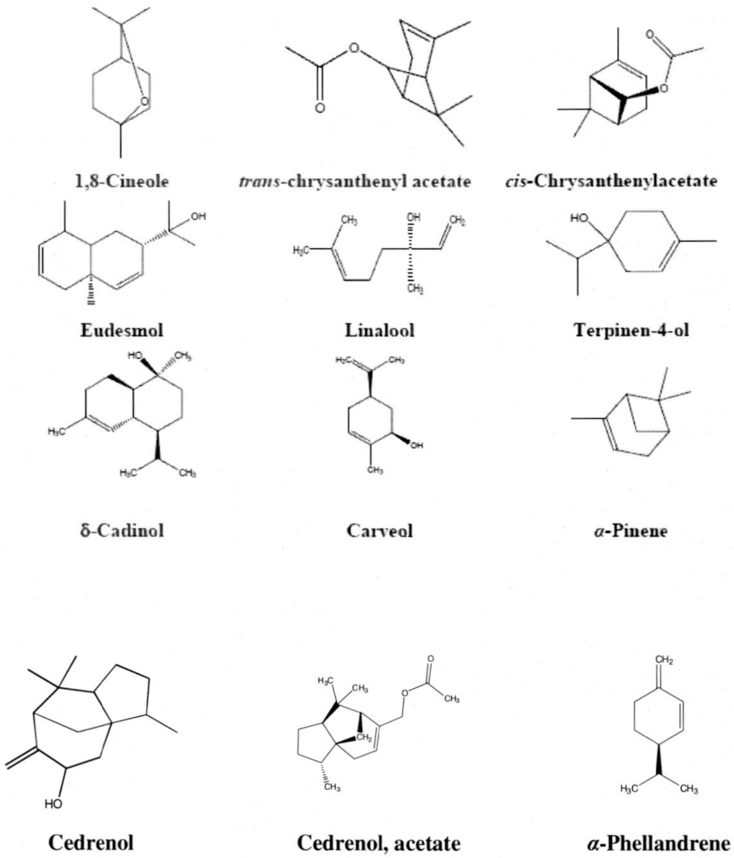

Figure 15. (Continued)

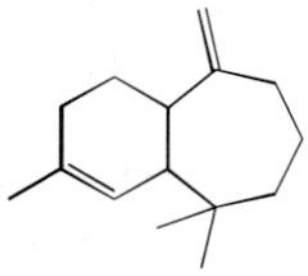

α-Himachalene

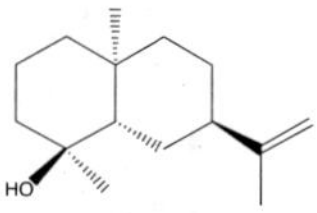

Intermedeol

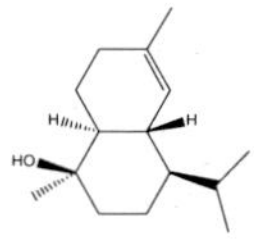

Cadinol

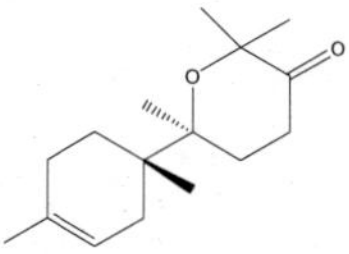

α-Bisabolone oxide

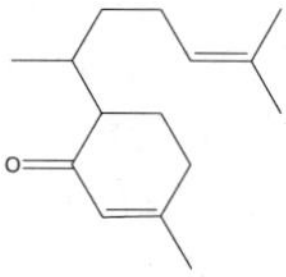

α-Oxobisabolene

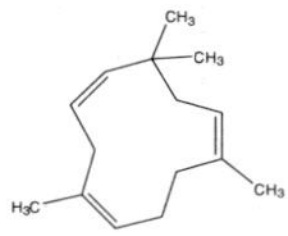

β-Humulene

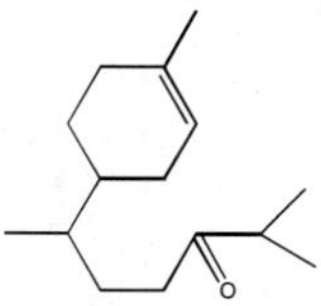

Bisabolone

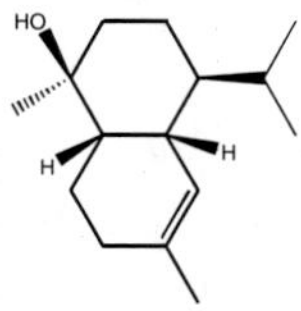

τ-Muurolol

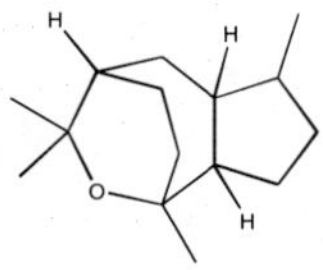

Kessane

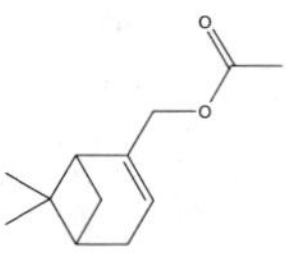

Myrtenyl acetate

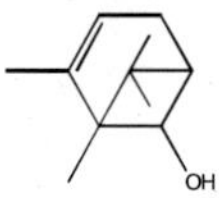

***cis*-Chrysanthenol**

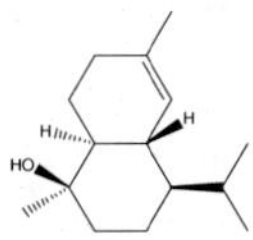

□-Cadinol

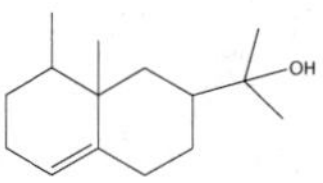

Valerianol

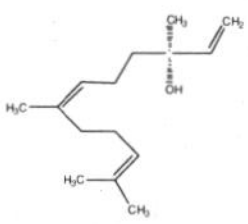

Nerolidol

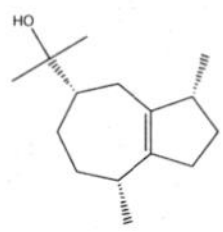

Guaiol

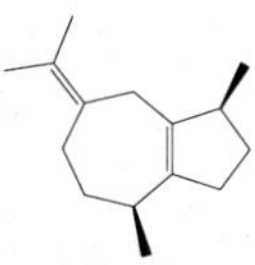

β-Guaiene

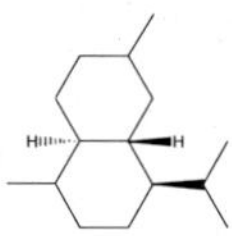

Cadinane

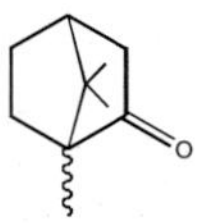

Camphor

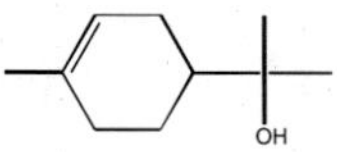

Terpineol

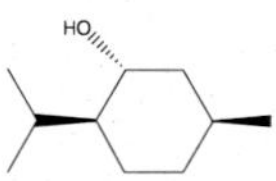

Isomenthol

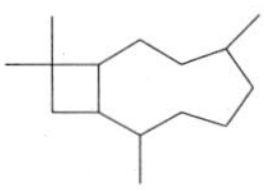

Caryophyllane

Figure 15. Structures of the main volatile terpenoids in EOs of endemic *Saharan Asteraceae.*

Various works carried out in Algeria and Morocco on the chemical analysis of essentialoils from the aerial parts *Warionia saharae,* indicates the richness of the essential in monoterpenes and sesquiterpenes, particularly the oxygeneted terpenoids. All studies indicate that the sesquiterpene lactone Eudesmol is the main compound of the EOs.

In the year 1985, Ramaut et al. [75] have identified only 3 constituents from *Warionia saharae* essential oils harvested from Morocco: Eudesmol (42.25%), linalool (8.63%) and nerolidol (17.26%). 25 years later, exactly in 2010, Cheriti et al. at the LPSO laboratory, carried out for the first time a preliminary analysis on the essential oils extracted from leaves of the Algerian specie, where they indicate the presence of cadinene (27.93%), guaiene (6.27%), and carophylane (3.36%) as the main compounds of the EOs [72]. In 2015, the same team completed the GC-FID and GC-MS analyzes of the essential oil obtained by hydrodistillation from leaves of *Warionea saharae*, reportedthat the Oxygen-containing terpene was the main group of constituents in EOs: Eudesmol (32.87%), isomenthol (6.27%), terpinyl butyrate (5.51%), trans-nerolidol (5.31%), linalool (4.99%), terpinen-4-ol (3.55%) and caryophyllane (3.36%) [73]. The Essential oils from the Moroccan *Warionia saharae* leaves were reported by several teams which indicated almost the same composition with eudesmol which varied between 23 and 50 % and nerolidol between 18 and 26%, in addition to the presence of linalool, guaiol, terpinen-4-ol and 1,8-cineol [74, 76-78].

Compared to the phytochemical analysis studies done on the EOs obtained from the Saharan species *Asteriscus graveolens* and *Warionia saharae,* there is a scarce work on the *Launaea arborescens* essential oils. Only one work dating since 2006, has been done by our team, where we found that the oil is rich in ester derivatives (dioctyl phthalate (38.6%) and decanoic acid, decyl ester (12.07%)) and aldehydes (16.09%). The terpenoid portion consisted of two oxygenated monoterpenes accounting for 7.24% of the oil [71].

Conclusion

It is well known that studying the secondary metabolites such as terpenoids of species flora is very important both for ecological, Phytochemical and pharmacological explorations. The importance of Sahara ecosystems for socioeconomic development is becoming increasingly recognized due to its ecological significance, biodiversity, particularity of local medicinal plants, and importance of medical heritage. Only few studies have been conducted on

Saharan traditional pharmacopeae, in the bioactivities assessment of local medicinal plants and in the exploration for new bioactive natural compounds.

The medicinal species from Saharan Asteraceae family (*Asteriscus graveolens*, *Launaea arborescens* and *Warionia saharae*) have great importance due to their ethnobotanics, phytochemistry and biological activities, and are promising source of various secondary metabolites including terpenoids (mono-sequiterpenes, triterpenes). Some of these isolates natural compounds especially sesquiterpene lactones have been found to exhibit various biological activities.

The present given informations on terpenoids from Saharan endemic plants can be the base for undertaking future research and to find new useful natural products. Thus, due to the chemical diversity and bioactivity of terpenoids, it is concluded that Saharan endemic Asteraceae species rich on these natural compounds need to be better investigated and to find adequate way to cultivation and preservation of local species in order to protect the Saharan ecosystem.

References

[1] Cheriti, A., (2020). "Endemic Asteraceae from Algerian Sahara: Potential medicinal value and chemical diversities, in *Asteraceae, Characteristics, Ecology and Distribution*edited by Millicent Tessier, 77-145, New York: Nova Science Publishers, USA. ISBN: 978-1-53616-632-3.

[2] Cheriti, A. and Belboukhari, N. (2015). Terpenoids of the Saharan medicinal plants *Launaea* Cass. Genus (Asteraceae) and their biological activities, in *Terpenoids and Squalene: Biosynthesis,Functions and Health Implications*, edited by Alanna B. Bates, 51- 70, Nova Science Publishers, New York, USA.

[3] Si-Yuan, P., Shu-Feng, Z., Si-Hua, G., Zhi-Ling, Y., Shuo-Feng, Z., Min-Ke, T., Jian-Ning, S., Dik-Lung, Ma.,Yi-Fan,H., Wang-Fun F. and Kam-Ming, K. (2013) "New Perspectives on How to Discover Drugs from Herbal Medicines: CAM's Outstanding Contribution to Modern Therapeutics" *Evidence-Based Complementary and Alternative Medicine*, doi: 10.1155/2013/627375.

[4] Bisset N., *Herbal drugs and phytopharmaceuticals. A handbook for practice on a scientific basis.*, (1994), Ed. Medpharm. Sc. Publishers, Sttutgarts-CRC Press, Boca Raton.

[5] Dobignard, A. and Chatelain, C. (2013). *Index synonymique de la flore d'afrique du nord* [*Synonymic index of the flora of North Africa*], Editions des conservatoire et jardin botaniques, Genève.

[6] Ibn al-Baytar, Dhiya al-Din. (1992). *Kitāb al-jāmiʿ li-mufradāt aladwiya wa al-aghdhiya* [*Compendium on Simple Medicaments and Foods*], Ed. Dar Kotob Elmia, Libanon.

[7] Bellakhdar, J. (1997). *La pharmacopée marocaine traditionnelle. Médecine arabe ancienne et savoirs populaires* [*Moroccan traditional pharmacopoeia. Ancient Arabic medicine and popular knowledge*], IBIS Press.

[8] a) Cheriti, A., Belboukhari, N. and Hacini, S., (2005) "Savoir traditionnel et valorisation des plantes médicinales du sud ouest Algérien" *An. Univ. Bec*.1,4-8. b) Belboukhri, N., Cheriti, A. and Roussel, C. (2007) "New Flavonoids from Butanol Extract of Launaea arborescens" *eCAM - Evidence-based Complementary and Alternative Medicine* -4(S1), 55. c) Cheriti, A., Belboukhari, N., Hacini, S.,(2001) *Annales Univ. de Bechar*,1. 4. d) Cheriti A., Belboukhari N., Hacini S., (2013), *PhytoChem & BioSub Journal*, 7(2), 52.

[9] Cheriti, A. (2000). *Medicinal Plants of the Bechar Region, Ethnopharmacologie studies*, CRSTRA Report, Algeria.

[10] Cheriti, A. (2019). *Limoniastrum feei* from the Algerian Sahara: Ethnopharmacology, Phytochemistry and the Pharmacological Importance, in *Plant sources and potential health benefits*, edited by Iman Ryan, 259-299, Nova Science Publishers, New York, USA.

[11] Buckingham J., Cooper C. M. and Purchase R. (2016). *Natural Products Desk Reference*; CRC Taylor &Francis, NY.

[12] Quezel, P. and Santa S. (1963). *Nouvelle flore de l'Algerie et desregions desrtiques méridionales* [*New flora of Algeria and southern deserts regions*], CNRS, Paris.

[13] Ozenda, P. (2004). *Flore et Végétation du Sahara* [*Flora and vegetation of the Sahara*], CNRS, Paris.

[14] Vicki A. F., Alfonso S., Tod F. S. and Randall B. (2009). *Systematics, Evolution, and Biogeography of Compositae*, IAPT, Vienna, Austria.

[15] Judd, W.S., Campbell, C.S., Kellogg, E., Stevens, P.F. (1999) *Plant Systematics:APhylogenetic Approach*. Sinauer, Sunderland, Massachusetts, USA.

[16] Wiklund, A. (1987) "The genus *Nauplius* (Asteraceae-Inuleae)", *Nord. J. Bot*. 7, 1-23.

[17] Cheriti, A., Saad, A., Belboukhri, N. and Ghezali, S. (2007) "The essential oil composition of *Bubonium graveolens* (Forssk.) Maire from the Algerian Sahara" *Flavour Fragr. J.*, *22*, 286-288.

[18] Greuter, W. (1997) "Save *Asteriscus*, sink *Nauplius* (Compositae)" *Fl. Medi.*, 7, 41-48.

[19] International Union of Conservation of Nature IUCN (2005) *A guide to medicinal plants in North Africa*, 1st Ed. IUCN Centre for Mediterranean Cooperation, Spain.

[20] Katinas, L., Tellería, M.C., Susanna, A. and Ortiz, S. (2008) "*Warionia* (Asteraceae): a relict genus of Cichorieae?" *Anales del Jardín Botánico de Madrid* 65(2), 367-381.

[21] Audissou, J.A. (1999) "*Warionia saharae*" *Brit. Cactus and Succulent J.* 17(3), 124-126.

[22] Bremer, K. (1994) *Asteraceae: Cladistics and classification*. Timber Press, Portland, Oregon.

[23] Cox, P.A. and Balick M.J. (1994) "The ethnobotanical approach to drug discovery" *Scientific American* 270(6), 82–87.

[24] World Health Organization (WHO) (2000) *General guidelines for methodologies on research and evaluation of traditional medicine*, WHO, Geneva.

[25] Wiart C. (2006). *Medicinal plants of the Asia-pacific drugs for the future,* World Scientific Publishing Co., Singapore.

[26] Cheriti, A., Rouissat, A., Sekkoum, K. and Balansard, G. (1995) "Plantes de la pharmacopée traditionelle dans la région d'El-Bayadh (Algérie)" *Fitoterapia* 66 (6), 525-538. ["Plants of the traditional pharmacopoeia in the region of El-Bayadh (Algeria)"]

[27] Cheriti, A., Belboukhari, N. and Hacini, S. (2013). "Ethnopharmacological survey and phytochemical screening of some medicinal Asteraceae from Algerian Sahara" *Phytochem & BioSub J.* 7 (2), 52-56.

[28] Cheriti, A. (2018). "Phytochemistry and biological activities of some endemic Saharan medicinal plants" *Paper presented at theInternational Days of Organic Chemistry*, Annaba, December 3-5.

[29] Aici, D., Cheriti, A., Bourmita, Y., Belboukhri, N. (2013) "Antimicrobial activity of essential oils of *Bubonium graveolens* (Forssk.)and *Anvillea radiata* (Coss.)" *Phytochem & Biosub J.* 7(3), 116-119.

[30] Melekmi, N., Saad, A., Belboukhari, N. and Cheriti, A. (2006) "Antimicrobial activity of the essential oil of *Bubonium graveolens*" *Ann.Univ. Bec.*, 2, 22-24.

[31] Chaib, F., Allali, H., Bennaceur, M. and Flamini, G. (2017) "Chemical Composition and Antimicrobial Activity of Essential Oils from the Aerial Parts of *Asteriscus graveolens* (Forssk.) Less. and *Pulicaria incisa* (Lam.) DC.: Two Asteraceae Herbs Growing Wild in the Hoggar" *Chem. Biodiversity* 14(8), doi:10.1002/cbdv.201700092.

[32] Alilou, H., Asdadi, A., Idrissi Hassani, L. M., González-Mas, C., Amparo Blázquez, M. and Akssira, M. (2014) "Antifungal and antioxidant activity of *Asteriscus graveolens* subsp. Odorus Essential Oil" *J. Nat. Sci. Res.* 4(10), 1-10.

[33] Znini, M., Cristofari, G., Majidi, L., Mazouz, H., Tomi, P., Paolini, J. and Costa, J. (2011) "Antifungal Activity of Essential Oil from *Asteriscus graveolens* against Postharvest Phytopathogenic Fungi in Apples," *Nat. Prod. Commun.*, *6*(11), 1763-1768.

[34] Znini, M., Cristofari, G., Majidi, L., Ansari, A., Bouyanzer, A., Paolini, J., Costa, J. and Hammouti, B. (2012) "Green approach to corrosion inhibition of mild steel by essential oil leaves of *Asteriscus graveolens* (Forssk.) in sulphuric acid medium" *Intern. J. Electrochem. Sci.,* 7, 3959-3981.

[35] Cristofari, G., Znini, M., Majidi, L., Mazouz, H., Tomi, P., Costa, J. and Paolini, J.(2012) "Chemical Diversity of Essential Oils from *Asteriscus graveolens* (Forssk.) Less: Identification of cis-8-Acetoxychrysanthenyl Acetate as a New Natural Component" *Chem. Biodivers.*, 9, 727-738.

[36] Aouissi, H., Gourine, N., Wang, H., Chen, X., Bombarda, I., Boudjeniba, M. and Yousfi, M. (2018) "Chemical composition, antioxidative, antimicrobial and anti-cancer activities of *Asteriscus graveolens* (Forssk) essential oil" *Oriental Pharm. Exper. Med.*, 18(3), 217–223.

[37] Boulenouar, N., Marouf, A., Cheriti, A. and Belboukhari, N. (2012). "Medicinal Plants Extracts as Source of Antifungal Agents against *Fusarium oxysporum* f. sp. *albedinis*" *J. Agr. Sci. Tech.* 14, 659-569.

[38] Keffous, F., Belboukhari, N., Djeradi, H., N.M.Rebizi, Cheriti, A., Sekkoum, K. and Aboul-Enein, H. Y. (2019) "Optimization of the reducing power for antioxidant activity test and application on two medicinal plants" *PhytoChem & BioSub J.* 13(2), 81-85.

[39] Cheriti, A., Belboukhari, M., Belboukhari, N. and Djeradi; H. (2012). "Phytochemical and biological studies on *Launaea* Cass. Genus (Asteraceae) from Algerian Sahara" *Current Topics in Phytochem* 11, 67-80.

[40] Belboukhari, N. and Cheriti, A. (2006) "Antibacterial and antifungal activities of crude extracts from *Launeae arborescens*" *Pak. J. Biolog. Sci.*, 9(1), 1-2.

[41] Bourmita, Y. (2014) *Toxicité comparée des extraits de quelques plantes spontanées de la région de Béchar chez des termines de type saharien* [*Comparative toxicity of extracts of some spontaneous plants from the Bechar region in Saharan-type termines*], Doctorat Diss., UKM-Ouargla.

[42] Bourmita, Y., Cheriti, A., Ould El Hadj, M. D., Mahmoudi, K. and Belboukhari, N. (2013). "Anti-termitic Activity of Aqueous Extracts from

Saharan Toxic Plants against *Anacanthotermes ochraceus,*" *J.Entomology* 10(4), 207-213.

[43] Salah Ramadan, B., Cheriti, A., Belboukhari, N. and Zaouani, M. (2008). "Evaluation de la toxicité de deux Asteraceae du Sahara Algérien: *Launaea Arbonescens* et *Bubonium gravoelens*" ["Evaluation of the toxicity of two Asteraceae from the Algerian Sahara: *Launaea Arbonescens* and *Bubonium gravoelens*"]. *An.Univ. Bec.* 4, 48-54.

[44] Watillon, C., Gaspar, T., Hofinger, M. and Ramaut, J.L. (1987) "La micropropagationde *Warionia saharae* Benth. & Coss." *Al Biruniya* 4: 35-38.

[45] Cheriti, A., Belboukhari, N., Bombarda, I. and Gaydou, E. (2010) "Antimicrobial Activity and Chemical Composition of the Essential Oil from the Sahara Endemic Medecinal Plant *Warionia saharae* (Asteraceae)" *Rev. Regions Arides*, 23(1), 103-107.

[46] Belboukhari, M., Cheriti, A. and Belboukhri, N. (2011) "Total phenolic content and *in vitro* antioxydant activity of extracts from the endemic medicinal plant *Warionia saharae*" *Nat. Prod.Ind. J.*, 7(3), 147-150.

[47] Amezouar, F., Badri, W., Hsaine, M., Bourhim, N. and Fougrach, H. (2012) Chemical Composition, Antioxidant and Antibacterial Activities of Leaves Essential Oil and Ethanolic Extract of Moroccan *Warionia saharae* Benth. & Coss." *J. App. Pharm. Sci.,* 2(5), 212-217.

[48] Znini, M., Cristofari, G., Majidi, L., El Harrak, A., Paolini, J. and Costa, J. (2013) "*In vitro* antifungal activity and chemical composition of *Warionia saharae* essential oil against 3 apple phytopathogenic fungi" *Food Sci. Biotechnol.* 22(S1), 113-119.

[49] Amezouar, F., Badri, W., Hsaine, M., Aksim M., Bourhim, N. and Fougrach, H.(2012) "subacute toxicity, anti-inflammatory and antioxidant activities ofethanolic extract of moroccan *warionia saharae* from Tata region" *Intern. J. Pharm. Pharm. Sci.* 4(5), 528-533.

[50] Berbaoui, H., Cheriti, A. and Ould El Hadj-Khelil, A. (2018) "Antibacterial Effect of Some Asteraceae of Southern Algeria on Nosocomial Strains of the Genus *Staphylococcus" Nat Prod Chem Res* 5: 302. doi:10.4172/2329-6836.1000302.

[51] Rouissat, L. (2017). *Etude des effets nématicides et molluscicides des extraits de quelques plantes sahariennes,* Doctorat Diss., UAB-Oran 1. [*Study of the nematicidal and molluscicidal effects of extracts from some Saharan plants*]

[52] Bergman M. E., Davis B., Phillips M. A., (2019) "Medically Useful Plant Terpenoids: Biosynthesis, Occurrence, and Mechanism of Action", *Molecules*, 24, 3961; doi:10.3390/molecules24213961.

[53] Perveen, S., Al-Taweel, A.M., Yusufoglu, H.S., Fawzy, G.A., Foudah, A. and Abdel-Kader, M.S. (2018) "Hepatoprotective and cytotoxic activities of Anvillea garcinii and isolation of four new secondary metabolites" *J. Nat. Med.* 72, 106–117.

[54] Belboukhari N. and Cheriti A. (2006)"Phytochemical Investigation of the Bioactive Extract from *Launeae arborescens". Pakistan Journal of Biological Sciences*, 9 (15): 2930-2932.

[55] Bitam, F., Ciavatta, M.L., Manzo, E., Dibi, A., Gavagnin, M. (2008) "Chemical characterisation of the terpenoid constituents of the Algerian plant Launaea arborescens" *Phytochemistry* 69 (17), 2984-2992.

[56] a) Abdel-Fattah, H., Zaghloul, A.M., Halim, A.F., Waight, E.S., *Egypt. J. Pharm. Sci*., (1990), 31, 81. b) Gherraf, N., El-Bassuony, A.A., Zellagui, A., Rhouati, S., Ahmed, A.A., Ouahrani, M.R., *Asian. J. Chem*., (2006), 18, 2348.

[57] Moumou, M., El Bouakher, A., Allouchi, H., El Hakmaoui, A., Benharref, A., Mathieu, V.,Guillaumet, G. and Akssira. M. (2014) "Synthesis and biological evaluation of 9*α*- and 9*β*-hydroxyamino-parthenolides as novel anticancer agents" *Bioorg. Med. Chem. Lett.* 24, 4014–4018.

[58] Mabry, T.J. and Bohlmann, F. (1977) Summary of the chemistry of the Compositae. In: Heywood, V.H. & al. (eds.), *The Biology andChemistry of the Compositae* 2, Academic Press, London.

[59] Hilmi, F., Gertsch, J., Bremner, P., Volovic, S., Heinrich, M., Sticher, O. and Heilmann, J. (2003) "Cytotoxic versus Anti-inflammatory effects in HeLa, Jurkat T and Human Peripheral Blood Cells caused by Guaianolide-Type sesquiterpene lactones" *Bioorg. Med. Chem.* 11, 3659-3663.

[60] Hilmi, F., Sticher, O. and Heilmann, J. (2002) "New Cytotoxic 6,7-cis and 6,7-trans configured Guaianolides from *Warionia saharae*" *J. Nat. Prod*., 65, 523-526.

[61] Hilmi, F., Sticher, O. and Heilmann, J. (2003) "New cytotoxic sesquiterpene lactonesfrom *Warionia saharae*. *Planta Med.* 69(5), 462-464.

[62] Akssira, M., Mellouki, F., Salhi, A., Alilou, H., Saouf, A., El Hanbali, F., Arteaga, J.F. and Barrerob, A.F. (2006) "Naupliolide, a sesquiterpene lactone with a novel tetracyclic skeleton from *Nauplius graveolens* subsp. Odorus" *Tetrahedron Lett.,* 47, 6719–6721.

[63] Djeradi H., Cheriti A., Rahmouni A. and Sekkoum Ka. (2021) *Docking study and DFT calculation on the inhibition of SARS-Cov2 by sesquiterpene lactones from Warionea saharae,* under publication.

[64] Loza-Tavera, H. (1999) "Monoterpenes In Essential Oils Biosynthesis And Properties," In F. Shahidi et al. (Eds.), *Chemicals Via Higher Plant Bioengineering,* Kluwer Academic /Plenum Publishers, New York.

[65] Aouissi, H., Gourine, N., Wang, H., Chen, X., Bombarda, I., Boudjeniba, M. and Yousfi, M. (2018) "Chemical composition, antioxidative, antimicrobial and anti-cancer activities of *Asteriscus graveolens* (Forssk) essential oil," *Oriental Pharm. Exper. Med.*, 18(3), 217–223.

[66] Chaib, F., Allali, H., Bennaceur, M. and Flamini, G. (2017) "Chemical Composition and Antimicrobial Activity of Essential Oils from the Aerial Parts of *Asteriscus graveolens* (Forssk.) Less. and *Pulicaria incisa* (Lam.) DC.: Two Asteraceae Herbs Growing Wild in the Hoggar" *Chem. Biodiversity*14(8), doi:10.1002/cbdv.201700092.

[67] Fahmy, H.S. (2003) "Chemical Composition and Antimicrobial Activity of the Volatile oil of *Nauplius graveolens* (Forssk.) Less." *J. Environ. Sci.*, 26, 307-317.

[68] Cristofari, G., Znini, M., Majidi, L., Mazouz, H., Tomi, P., Costa, J. and Paolini, J.(2012) "Chemical Diversity of Essential Oils from *Asteriscus graveolens* (Forssk.) Less: Identification of cis-8-Acetoxychrysanthenyl Acetate as a New Natural Component" *Chem. Biodivers.*, 9, 727-738.

[69] Znini, M., Cristofari, G., Majidi, L., Mazouz, H., Tomi, P., Paolini, J. and Costa, J. (2011) "Antifungal Activity of Essential Oil from *Asteriscus graveolens* against Postharvest Phytopathogenic Fungi in Apples," *Nat. Prod. Commun.*, *6*(11), 1763-1768.

[70] Alilou, H., Asdadi, A., Idrissi Hassani, L. M., González-Mas, C., Amparo Blázquez, M. and Akssira, M. (2014) "Antifungal and antioxidant activity of *Asteriscus graveolens* subsp. Odorus Essential Oil" *J. Nat. Sci. Res.* 4(10), 1-10.

[71] Cheriti A., Saad A., Belboukhari N. et Ghezali S. (2006) "Chemical composition of the essential oil of *Launeae arboresens* from algerian Sahara." *Chemistry of Natural Compunds.* 42 (3):360-361 70.

[72] Cheriti, A., Belboukhari, N., Bombarda, I. and Gaydou, E. (2010) "Antimicrobial Activity and Chemical Composition of the Essential Oil from the Sahara Endemic Medecinal Plant *Warionia saharae* (Asteraceae)" *Rev. Regions Arides*, 23(1), 103-107.

[73] Yakoubi, M., Belboukhari, N., Sekkoum, K., Cheriti, A., Bouchekara, M.and Aboul-Enein, H. Y. (2015) "The Chemical Composition of Essential Oil of *Warionea saharae* Benth & Coss.(Fam. Asteraceae) from Southern Algeria" *Current Bioactive Compounds*, 11, 116-121.

[74] Essaqui, A., Elamrani, A., J. A. Cayuela and Benaissa, M. (2007) "Chemical Composition of the Essential Oil of Warionia saharae from Morocco," *J. Ess. Oil Bear. Plants* 10(3), 241 -246.

[75] Ramaut, J. L., Hofimger, M., Dimbi, R., Corvisier, M. and Lewalle, J. (1985) "Main Constituents of the Essential Oil of *Warionia Saharae* Benth and Coss." *Chromatographia* 20, 193-194.

[76] Znini, M., Cristofari, G., Majidi, L., El Harrak, A., Paolini, J. and Costa, J. (2013) "*In vitro* antifungal activity and chemical composition of *Warionia saharae* essential oil against 3 apple phytopathogenic fungi" *Food Sci. Biotechnol.* 22(S1), 113-119.

[77] Amezouar, F., Badri, W., Hsaine, M., Bourhim, N. and Fougrach, H. (2012) Chemical Composition, Antioxidant and Antibacterial Activities of Leaves Essential Oil and Ethanolic Extract of Moroccan *Warionia saharae* Benth. & Coss." *J. App. Pharm. Sci., 2*(5), 212-217.

[78] Sellam, K., Ramchoun, M., Alem, C., Khallouki, F., El Moualij, B. and El Rhaffari1, L. (2012) Chemical composition antioxidant and antimicrobial activities of essential oil of warionia saharae from oases of morocco, In: *Gas Chromatography – Biochemicals, Narcotics and Essential Oils*; Bekir Salih, Ömür Çelikb çak, Eds.; InTech, pp. 213-22.

[79] Said, M.E., Bombarda, I., Naubron, J.V., Vanloot, P.J. M., Cheriti, A., Dupuy, N. and Roussel, C. (2017) "Isolation of the major chiral compounds from *Bubonium graveolens* essential oil by HPLC and absolute configuration determination by VCD" *Chirality* 29, 70-79.

Chapter 3

Keggin Heteropolyacid Salts-Catalyzed Terpenes Epoxide Synthesis

Marcio Jose da Silva* **and Castelo Bandane Vilanculo**
Chemistry Department, Federal University of Vicosa,
Vicosa, Minas Gerais State, Brazil

Abstract

Terpenic alcohols are oxygenated derivatives from monoterpenes and can be easily obtained from natural raw materials such as extracts of wood and plants, and essential oils. Several oxidation products obtained from terpenic alcohols have pleasant organoleptic properties and are used as an ingredient in the synthesis of highly valuable compounds such as flavours, drugs, agrochemicals, and other fine chemicals. Controlling the reaction selectivity is one of the main challenges of these processes, mainly because these substrates can give carbonylic derivatives, isomers, and other products. Another problem in epoxidation reactions involving terpenic alcohols or other olefins is the use of stoichiometric quantities of toxic oxidants, which generate residues that should be disposed of into the environment. To circumvent this drawback, a green oxidant such as hydrogen peroxide can be used, however, it is always required the presence of a catalyst to do feasible these reactions. Noble metal catalysts are efficient in this process, nonetheless, their high price is a great disadvantage. An attractive alternative is the use of Keggin

* Corresponding Author's E-mail: silvamj2003@ufv.br

In: Terpenes and Terpenoids
Editor: Charles A. Davies
ISBN: 978-1-68507-559-0

heteropolyacid (i.e., HPAs) or their derivatives salts as catalysts. In this work, we would like to highlight recent advances achieved in the development of processes for epoxides production from terpenic alcohols in Keggin heteropolyacid salts-catalyzed reactions using hydrogen peroxide as oxidant. Keggin HPAs are metal-oxygen clusters highly versatile as homogeneous or heterogeneous catalysts. On the other hand, hydrogen peroxide is a green oxidant that produces only water as a by-product. Therefore, we have developed a catalyst system using Keggin HPA salts and hydrogen peroxide to epoxidize terpenic alcohols (i.e., geraniol, nerol, linalool, α-terpineol, and β-citronellol). We will pay special attention to assessing the effects of main reaction parameters such as concentration, temperature, and stoichiometry of reactants. A comparison with traditional acid catalysts used in these reactions was also performed. Advances in the development of a new, efficient, and environmentally benign catalytic process to convert terpenic alcohols to epoxide were described. The authors hope that this work can significantly contribute to improving this important research field.

Keywords: Keggin heteropolyacids, lacunar heteropolyacid salts, metal-exchanged heteropolyacid salts, terpenic alcohols, epoxidation, hydrogen peroxide

Introduction

Oxidation of Terpene Alcohols

Monoterpenes are renewable raw materials used to synthesize perfumes, agrochemicals, and drugs intermediates (Gallezoti, 2012). The synthesis of the oxygenated products derived from monoterpenes has great interest due to its use in the production of building blocks for the synthesis of medicines, agrochemicals, fragrance additives and perfumes (Tsolakis et al., 2019; Da Silva et al., 2015).

Oxidative processes become highly attractive when green oxidants are involved. Moreover, the presence of an active and recyclable catalyst is an essential condition to the process to be environmentally friendly (Sadri et al., 2014). The catalytic processes must achieve high activity and selectivity toward the desired product, with a minimum generation of waste and effluents (Guo et al., 2014).

In this work, oxidative processes where substrates of renewable origin such as the terpenic alcohols nerol, linalool, geraniol, and borneol (Figure 1)

were developed, using HPA salts as the catalysts, and hydrogen peroxide as the oxidant.

Besides the epoxides, oxygenated derivatives of the terpene alcohols likewise aldehydes, ketones and esters have attractive olfactory properties and are widely used as ingredients for perfumery and fragrance industries (De Oliveira et al., 2009). Particularly, the oxidation of terpene alcohols becomes even more interesting when it is carried out with a non-polluting oxidant, such as hydrogen peroxide, which is non-flammable, easy to handle, cheap, and atom efficient (De Oliveira et al., 2009; Da Silva et al., 2009).

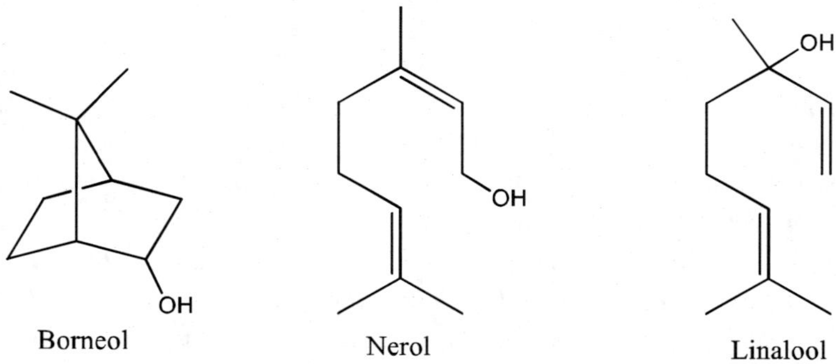

Figure 1. Main terpene alcohols used in this work.

The greatest challenge in the oxidation of terpene alcohols is that they can have double bonds with different reactivity, which sometimes undergo a steric hindrance, thus being difficult to be epoxidized (Da Silva et al., 2004). There are scarce examples of metal catalysts successfully used in the epoxidation of terpene alcohols with hydrogen peroxide (Da Silva et al., 2001; Maksimchuk et al., 2005).

Terpenic alcohols are an abundant, renewable, and attractive raw material for the synthesis of various fine chemical products (Da Silva et al., 2018). Linalool is tertiary alcohol whose cyclization derivatives and epoxides are highly valuable because they are chiral building blocks for drug synthesis (Vilanculo and Da Silva, 2020).

On the other hand, terpenic alcohols can be also converted to ethers, which are attractive raw materials for the synthesis of fine chemicals, drugs and agrochemicals (Da Silva et al., 2018). Ethers derived from tetrahydropyran and tetrahydrofuran (i.e., linalool oxides) with the general structure of type 2 and 3 (Scheme 1) are natural compounds used in the synthesis of biologically active compounds, with potential application in medicine. In addition, these ethers are fragrance and flavour ingredients (Serra and Simeis, 2018; Nasir et al., 2014).

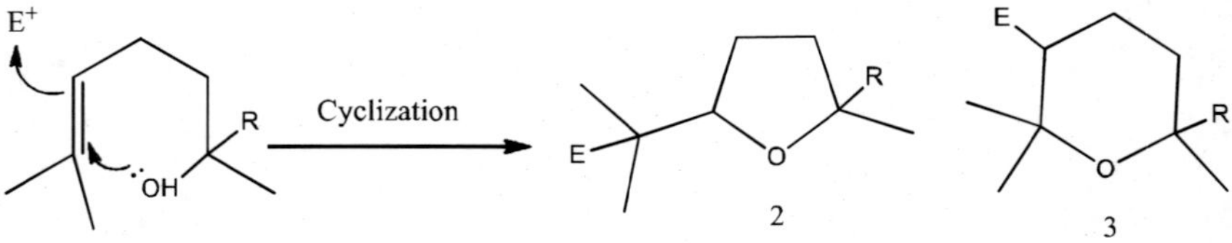

1 2 3

E = H, OH, Vinyl, aril, alkyl, halide

Scheme 1. Cyclization of unsaturated terpene alcohol.

The synthesis of these ethers is generally based on the cyclization of terpene alcohols (i.e., type 1), and follows two main pathways, both involving the intramolecular nucleophilic addition of a hydroxyl group to a terminal prenyl group (Boivin, 1987, Nicolaou et al., 1989).

Commonly, this synthesis requires first a preliminary step to activate the double bond through its transformation into epoxide or halide derivatives, which cyclize generating the corresponding ether derivatives, with an additional functional group of hydroxyl or halogen (Nasir et al., 2014). Ali et al. described a synthetic route of trans-(+)-linalool oxide, using the substrate 2-methylen-6-ol-heptene as starting material, and $NaMnO_4$ as an oxidant (Ali et al., 2014). Other stoichiometric oxidants such as $KMnO_4$ or NaClO are commonly used for the synthesis of linalool oxides. However, these oxidants have a low atomic economy and generate a large amount of waste.

Alternatively, hydrogen peroxide can be used as the oxidant; it is a cheap reactant, easy to handle, with minimal environmental impact since that generates water as the only by-product. Hydrogen peroxide has a good atomic efficiency (ca. 47%), it is non-flammable and very effective in liquid phase oxidation reactions (Bryliakov et al., 2017). However, its activation requires the use of a metal catalyst, which should promote a faster transfer of the Oxygen atom from the oxidant to the substrate and increase the reaction selectivity (Guo et al., 2014; Duprez and Cavani, 2014; Zhang et al., 2019).

Catalysis as a Tool to Add Value to the Terpenes

Since the last century that catalytic processes have been recognized worldwide as a technology that has quickly grown. Environmental concerns have motivated the development of new catalysts, with a constant search for sustainable technologies and low energy costs, whose purpose is to ensure survival and well-being in the modern world. In this sense, the chemical industry has adopted various attitudes to minimize the generation of waste and effluents, such as developing processes where both catalyst and reactants have a low environmental impact. Moreover, renewable feedstock has been preferred whereas the fossil origin substrates (Da Silva et al., 2016; Da Silva et al., 2018; Viana et al., 2018). The so-called "Green Chemistry" seeks to make the raw material transformation processes cleaner, using non-toxic reagents, to generate smaller amounts of waste and reduce its environmental impact.

Catalysis science is devoted to the study of the factors that affect the reaction rates, and which of them can draw the selectivity toward the desired product. The catalyst is a substance that is added to the reaction system and that changes the reaction rate until it reaches chemical equilibrium (Astruc, 2007; Bernardo-Gusmão et al., 2017). Nevertheless, the catalyst does not affect the Gibbs energy of the global reaction, since this is a state function. Since it presents a purely kinetic effect, it accelerates only thermodynamically favourable reactions (Oro & Sola, 2000).

The catalyst participates in a chemical reaction without being consumed at their end (Norskov et al., 2014). It provides an alternative pathway for the reaction, where the activation energy is lower, allowing the catalytic reaction to become faster and more selective (Oro&Solo, 2000; Chorkhendorff & Niemantsverdriet, 2003).

More than 80% of industrial chemicals are catalytically produced. A catalyst has an important property (activity), which can be expressed in different ways (Norskov et al., 2014). The number of rotations (Turnover number-TON), expresses the number of mol of products obtained per mol of the catalyst and provides a measure of the activity of the catalyst. The speed of catalytic action can be expressed by the rotation frequency (TOF), which is defined as the number of moles of product per unit of time, divided by the number of mols of catalyst. An efficient catalyst has both high TON and TOF (Oro & Solo, 2000).

Heteropolyacid Catalysts

Different solid catalysts have been widely developed to be used in terpenic alcohol oxidation reactions (Dusi et al., 2000; De Vos et al., 2003). Niobium

(Batalha et al., 2020), Tungsten (Wang & Yamamoto, 2014) Titanium (Kumar et al., 1995) and mixed metal oxides (Marin-Astorga et al., 2012; Somma & Strukul, 2004) are just a few examples of catalysts used in oxidation and/or epoxidation reactions of terpenic alcohols with hydrogen peroxide.

Keggin Heteropolyacids (HPAs) are considered environmentally benign catalysts and are a good alternative to the common metal-containing catalysts (Bond et al., 2012; Haider Et al., 2012; Lopez et al., 2012; Lv et al., 2013; Mekki-Berrada et al., 2013; Omwoma et al., 2015; Da Silva, 2016; Da Silva et al; 2016; Coronel & Da Silva, 2018) (Figure 1).

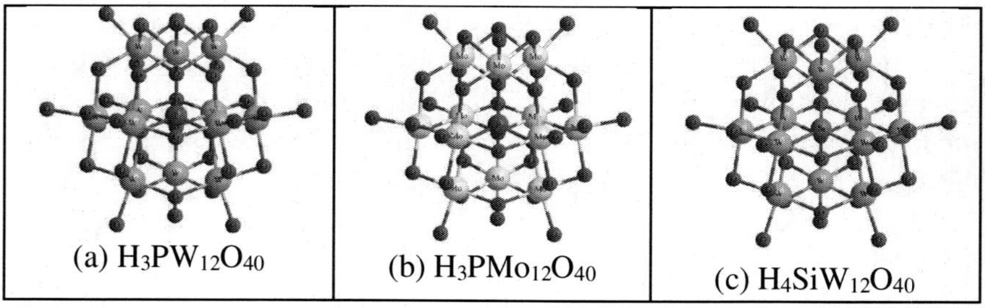

Figure 2. Keggin heteropolyacids: phosphotungstic (a), phosphomolybdic (b), and silicotungstic (c) acids.

The Keggin anion has a heteroatom (P or Si), linked to 4 Oxygen atoms, forming a central tetrahedron. This tetrahedron is surrounded by 12 MO_6 octahedra, connected by sharing vertices and edges. The 12 octahedra have 4 groups of 3 shared octahedra, forming M_3O_{13} triads, which have a common apex of Oxygen attached to the central atom. (Kourasi et al., 2014; Narkhede et al., 2015). In addition to exhibiting a strong Brønsted acidity, suitable for acid-catalyzed organic reactions, POMs have redox properties, which allow a rapid and reversible control of the oxidation number of the metal. This characteristic is essential for that POMs can be efficient catalysts in oxidation reactions of alkanes, aromatic compounds, olefins, and alcohols (Farsani & Yadollahi, 2014; Song et al., 2016).

Keggin-type HPAs have Tungsten or Molybdenum as "addendum" atoms in their anions and are therefore potentially active catalysts in oxidation reactions (Cavani, 1998; Casuscelli et al., 2004; Amitouche et al., 2018). However, HPAs are strong Brønsted acids, a characteristic that *can promote* competitive reactions and reduce the selectivity of the oxidation products (De Meireles et al., 2011; Zheng Et al., 2018).

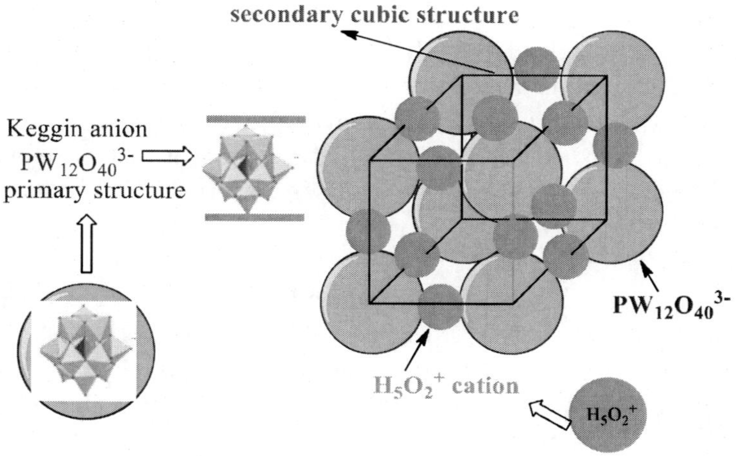

Figure 3. Primary and secondary structures of Keggin HPAs (adapted from Batalha et al., 2020).

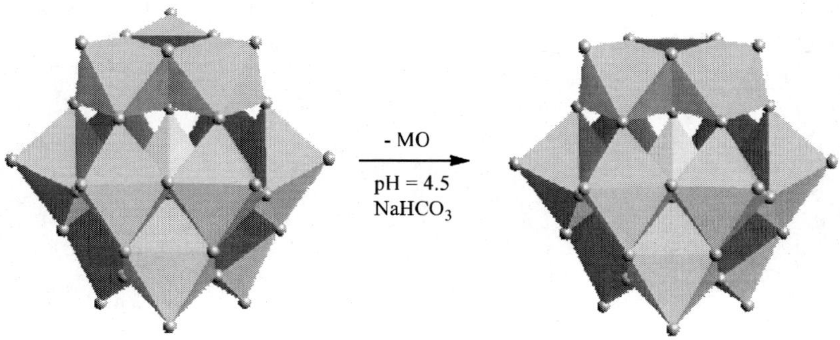

Figure 4. Conversion of Keggin heteropolyanion (i.e., $PW_{12}O_{40}^{-3}$ or $PMo_{12}O_{40}^{-3}$) to lacunar Keggin heteropolyanion (i.e., $PW_{11}O_{39}^{-7}$ or $PMo_{11}O_{39}^{-7}$) (Da Silva et al., 2018).

To overcome this disadvantage, an approach that is still little explored is to convert Keggin's HPAs to neutral salts, exchanging their acidic protons for metal cations (Karcz et al., 2017; Da Silva & De Oliveira, 2018). This modification changes the physical properties of the Keggin HPAs, such as the surface area, porosity, and solubility, but keeps untouched the heteropolyanion structure (Figure 3).

Furthermore, the removal of the MO unit from heteropolyanions (i.e., WO or MoO) results in a lacunar salt, which may be more active than their catalytic precursor, which has a saturated heteropolyanion (Kamata et al., 2007; Li et al., 2015; Coronel & Da Silva, 2018; Da Silva et al., 2018).

Lacunar POMs belong to the class of inorganic binders, with special properties such as high softness, and good oxidative stability (Borras-Almenar et al., 2003; Narkhede et al., 2015). The controlled treatment of POMs solutions with a base (i.e., $NaHCO_3$), leads to the formation of lacunar POMs, where one or more addendum atoms (i.e., W, Mo, etc.) can be removed from the structure, together with the Oxygen atoms (POPE, 1983). Removal of one or two MO units from the saturated anion $[XMo_{12}O_{40}]^{n-}$ POM leads to the formation of mono or di-lacunar POMs, $[XMo(V)_{11}O_{39}]^{(n+4)}$- and $[XMo(V)_{10}O_{36}]^{(n+5)-}$, where X is the heteroatom and M is the addendum atom.

Literature has described only scarce examples of the oxy-functionalization of monoterpenes or terpene alcohols, using monolacunar POMs as catalysts. Balula et al. (2013) studied the epoxidation of geraniol and limonene, using monolacunar phosphotungstate salt ($PW_{11}M$) and sandwich-type salt $M_4(PW_9)_2$ as catalysts, and hydrogen peroxide as an oxidant. Those authors verified that when Co(II) or Fe(III) were the metal cations, a 100% conversion of geraniol and limonene were achieved after 1.5 and 3 h of reaction, respectively.

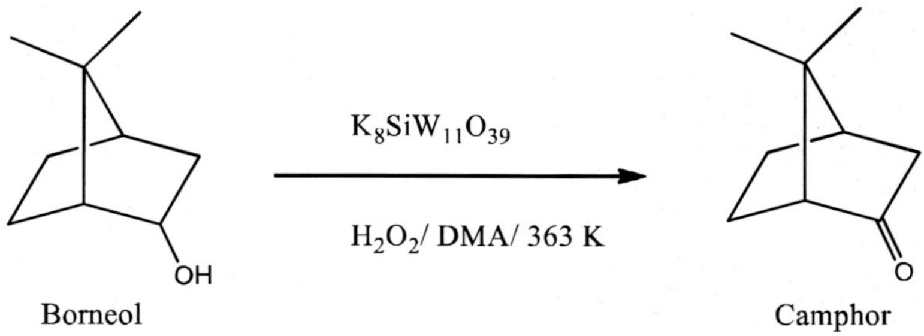

Scheme 2. $K_8SiW_{11}O_{39}$-catalyzed oxidation of borneol with hydrogen peroxide to camphor (Da Silva et al., 2018).

Da Silva et al. recently described reactions of oxidation of borneol with H_2O_2 in dimethylacetamide solutions, using $K_8SiW_{11}O_{39}$ as the catalyst, and verified that a high conversion and selectivity (ca. 100%) toward camphor were simultaneously reached (Da Silva et al., 2018).

Lacunar Cesium salts were synthesized starting from three Keggin HPAs (i.e., $H_4SiW_{12}O_{40}$, $H_3PW_{12}O_{40}$, and $H_3PMo_{12}O_{40}$) and evaluated in the oxidation of terpenic alcohols (Batalha et al., 2020). Among them, the lacunar Cesium silicotung state salt (i.e., $Cs_8SiW_{11}O_{39}$) was the most active in the oxidation of nerol with hydrogen peroxide, reaching at the same time 90% of conversion and selectivity to epoxy-nerol (Scheme 3).

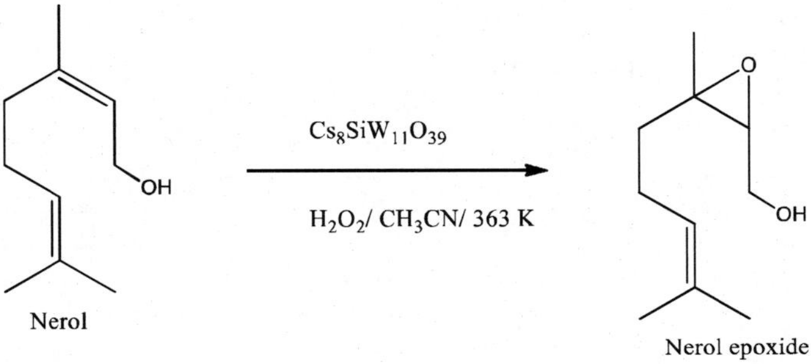

Scheme 3. $Cs_8SiW_{11}O_{39}$-catalyzed epoxidation of nerol with hydrogen peroxide (Batalha et al., 2020).

Lacunar Sodium heteropoly salts were synthesized and tested in the oxidation reactions of linalool with hydrogen peroxide (Vilanculo et al., 2020). In this case, surprisingly, two cyclization products were also obtained (Scheme 4).

Scheme 4. Oxidation of linalool by H2O2 catalyzed by Na7PW11O39.

The effect of heteropolyanion was also assessed and the main results are summarized in Table 1. Note that cis/trans 2-(5-methyl-5-vinyltetrahydrofuran-2-ylpropan-2-ol) (1a) and cis/trans 2,2,6-trimethyl-6-vinyltetrahydro-2H-pyran-3-ol (1b) were the main products, with 66 and 27% selectivity, respectively, while diepoxide (1c) was obtained with only 7% selectivity (Scheme 4).

The results indicated that this catalytic system offers a good route for the synthesis of linalool oxides. The formation of the tetrahydrofuran derivative (1a), with a five-membered ring, was more favourable and occurred in all reactions, except when the $Na_7PMo_{11}O_{39}$ was the catalyst. Probably, the tetrahydrofuran derivative is the kinetic product, so it is formed more quickly than the six-membered ring product.

Table 1. Effect of heteropoly salt catalyst on the linalool oxidation with hydrogen peroxide

Exp.	Catalyst	Conversion (%)	Selectivity (%)		
			1a	1b	1c
1	$Na_7PMo_{11}O_{39}$	32	24	17	59
2	$Na_7PW_{11}O_{39}$	100	66	27	7
3	$Na_8SiW_{11}O_{39}$	11	60	19	21
4	$Na_3PW_{12}O_{40}$	3	31	0	0

Based on experimental data previously reported in the literature, it was possible to make some insights into the mechanism for the synthesis of linalool oxides.

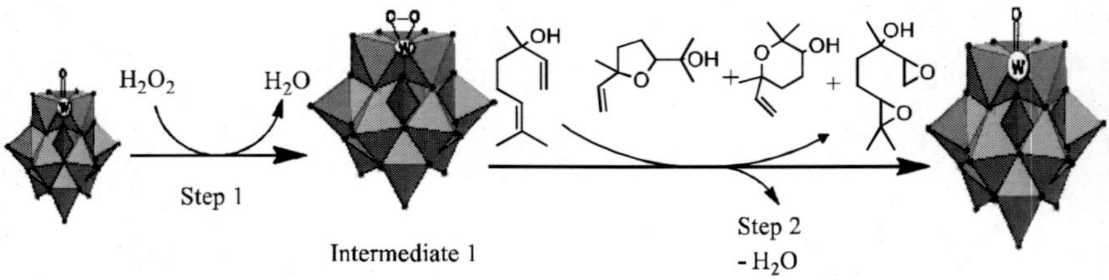

Scheme 5. Presumed reaction path for the oxidation of linalool by hydrogen peroxide catalyzed by $Na_7PW_{11}O_{39}$ (adapted from Da Silva et al., 2018; Viana et al., 2018; Vilanculo et al, 2020).

An excess of peroxide is necessary to peroxidize the lacunar Sodium phosphotungstate catalyst, as depicted in Scheme 5. The literature has reported the same phenomena for oxidation reactions with H_2O_2 in the presence of Tungsten catalysts (Da Silva et al., 2018; Viana et al., 2018).

Thus, after the formation of intermediate 1, it then transfers the Oxygen atom to the double bond of linalool. This transformation firstly results in an epoxide, which can be easily cyclized to the corresponding ether derivatives (i.e., linalool oxides; 1a and 1b) containing an additional hydroxyl group.

Therefore, it suggests that the epoxide formation is a key aspect for the reaction to proceed. The intramolecular nucleophilic attack of the hydroxyl group belonging to the linalool epoxide is the essential step that governs the selectivity of the reaction (Vilanculo et al., 2020). After the epoxide formation, its oxirane ring can be opened when the pairs of electrons from the hydroxyl group attack the less hindered electrophilic Carbon atom.

This intramolecular cyclization step gives the tetrahydrofuran after a transfer of the proton to anion. On the other hand, when pairs of electrons from the hydroxyl group attack the most hindered electrophilic Carbon atom, the tetrahydropyran is the product mainly obtained. Note that with epoxides, regioselectivity is not that simple, because, even with acidic catalysts, the replacement of the A_ND_N type in a primary centre is very fast (Clayden et al., 2000).

Table 2. Effect of catalyst concentration on linalool oxidation reactions by hydrogen peroxide catalyzed by $Na_7PW_{11}O_{39}$

Run	Concentracion (mol %)	Conversion (%)	Selectivity (%)		
			2a	2b	2c
1	0.083	59	71	25	4
2	0.165	90	73	24	3
3	0.250	97	72	25	3
4	0.330	100	66	27	7
5	0.500	100	69	27	3

The effect of the amount of H_2O_2 was investigated, and it was found that while the reaction conversions were strongly affected, the reaction selectivity remained almost unchanged. After varying the catalyst concentration, it was observed that the same effect was triggered (Table 2).

The same catalyst was also assessed in the nerol oxidation (2) with hydrogen peroxide (Vilnaculo et al, 2020).

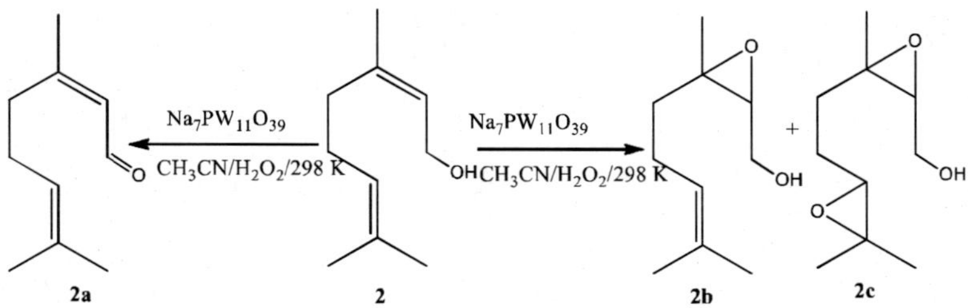

Scheme 6. $Na_7PW_{11}O_{39}$-catalyzed oxidation of nerol with hydrogen peroxide.

Neraldehyde (2a), nerol epoxide (2b) and diepoxide (2c) were obtained as the main products in the nerol oxidation reaction (Scheme 6). Nerol epoxide was the major product (ca. 91% selectivity), while the aldehyde selectivity was only 5% (Table 2). This system suggests that the double bond of allyl alcohol is easily activated, leading to the formation of epoxide as the main product (Table 3).

Table 3. Effect of heteropoly salt catalyst on the nerol oxidation with hydrogen peroxide

Run	Catalyst	Conversion (%)	Selectivity (%)			
			1a	1b	1c	Nerol peroxide
1	$Na_7PW_{11}O_{39}$	95	5	91	2	2
2	$Na_8SiW_{11}O_{39}$	71	6	89	0	5
3	$Na_7PMo_{11}O_{39}$	27	12	82	0	6
4	$Na_3PW_{12}O_{40}$	3	14	25	0	61

When the saturated Sodium phosphotungstate salt (i.e., $Na_3PW_{12}O_{40}$) was the catalyst, only 3% of conversion was achieved and, under these experimental conditions, a low selectivity of 2a and 2b were observed (14 and 25%, respectively). A trace of 2c was detected, suggesting that alkyl peroxides were formed in large amounts (Table 3). This result suggests that the reaction is promoted by lacunar salts, especially the lacunar phosphotungstate.

The impact of catalyst load was also evaluated (Table 4). When the reactions were carried out in the range of 0.0165 to 0.166 mol% of the catalyst, a greater amount of undetected products was observed, meaning that in this range of concentration, part of the substrate was still as alkyl intermediates peroxides.

Table 4. Effect of $Na_7PW_{11}O_{39}$ catalyst concentration on the nerol oxidation reactions with hydrogen peroxide

Run	Catalyst mol (%)	Conversion (%)	Selectivity (%)			
			1a	1b	1c	Nerol peroxide
1	0.0165	63	7	93	0	0
2	0.033	63	4	60	0	37
3	0.083	83	4	59	0	36
4	0.165	87	4	69	0	27
5	0.33	95	5	92	1	2

It also means that, even at an equimolar amount of nerol to oxidant, a minimal amount of catalyst is needed to decompose the alkyl peroxide intermediates to oxidation products. Thus, we decided to increase the amount of catalyst in the reaction. When 0.33 mol% was used, higher substrate conversion and epoxide selectivity were achieved and the alkyl peroxide intermediates were significantly reduced.

Another modification that can improve the activity of Keggin HPAs is to exchange one or more Molybdenum atoms with Vanadium atoms. This modification makes faster the oxidation-reduction steps, resulting in improved activity and higher selectivity in oxidation reactions (Kozhevnikov, 1997; Lee et al., 2001; Barteaua et al., 2006; Mizuno & Kamata, 2011; Jing et al., 2014).

The insertion of a Vanadium atom into the Keggin heteropolyanion has a significant effect on the catalyst performance. The replacement of a Mo atom by a V atom in the catalyst's primary structure accelerates both the reduction and reoxidation steps, which results in an improvement in activity and selectivity (JING et al., 2014). Scheme 7 describes the synthesis of $H_4PMo_{11}VO_{40}$ and their Sodium salt $Na_4PMo_{11}VO_{40}$.

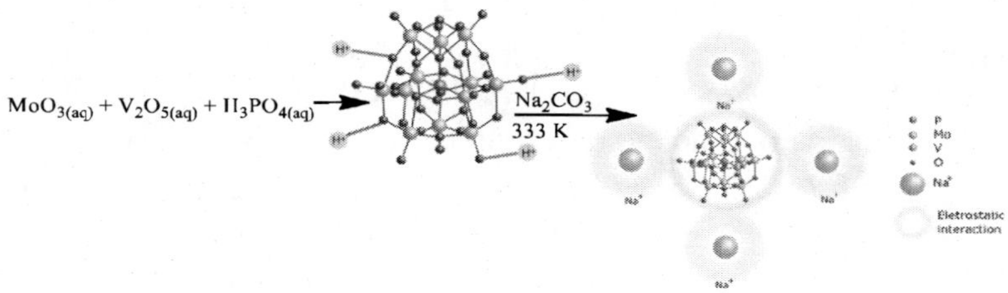

Scheme 7. Synthesis of $H_4PMo_{11}VO_{40}$ and their Sodium salt $Na_4PMo_{11}VO_{40}$.

Recently, nerol oxidation was performed using $Na_4PMo_{11}VO_{40}$ as the catalyst (Vilanculo and Da Silva, 2021). The activity of catalysts followed the trends $Na_4PMo_{12}VO_{40}$ > $H_3PMo_{12}O_{40}$ > $Na_3PMo_{12}O_{40}$. It suggests that Vanadium doping had a beneficial effect on the performance of the Sodium phosphomolybdate catalyst. Besides the higher initial rate and conversion, this catalyst provided the greatest selectivity for the oxidation products, with nerol epoxide being the main product (Scheme 5). Nerol diepoxide and neraldehyde were the secondary products (Scheme 2-9).

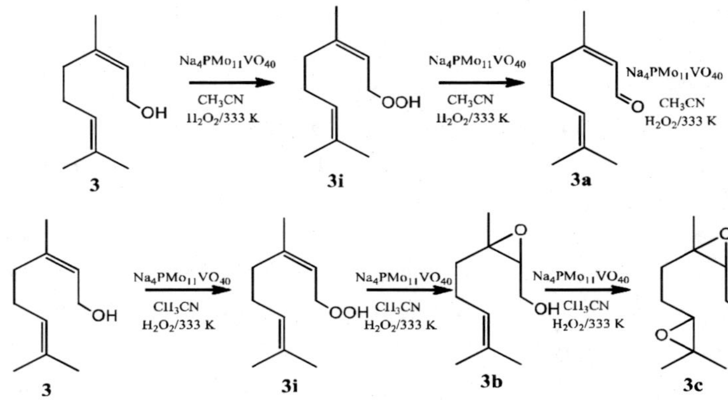

Scheme 8. $Na_4PMo_{12}VO_{40}$-catalyzed oxidation of nerol with hydrogen peroxide.

The undoped catalyst had the lowest conversion (ca. 73%), it had the lowest selectivity toward epoxide nerol. Furthermore, in this reaction, there was verified the greatest formation of alkyl peroxides, which are possible reaction intermediates. These compounds were not detected by GC-FID analysis but were calculated using the mass balance of the reactions (Scheme 8, product 3d).

Note that aldehyde (3a), nerol epoxide (3b) and diepoxide (3c) were obtained in the nerol oxidation reaction (Scheme 8). Nerol epoxide was the major product (ca. 96% selectivity); conversely, neraldehyde had low selectivity. Similar to what was observed in previous studies using $Na_7PW_{11}O_{39}$ as a catalyst, the result obtained in this system suggests that the allylic alcohol double bond is easily activated, leading to the formation of the epoxide as the main product (Vilanculo et al., 2020).

Therefore, it is possible to conclude that the Vanadium may be part of the active site in the salt structure, making it a potential catalyst in the oxidation of terpenic alcohols. The highest conversion of nerol was achieved (ca. 96%) when $Na_4PMo_{11}VO_{40}$ was the catalyst.

When the Sodium phosphomolybdate salt ($Na_3PMo_{12}O_{40}$) was the catalyst, three products were obtained in the oxidation of nerol by hydrogen peroxide: neraldehyde (3a), nerol epoxide (3b) and diepoxide (3c). However, the conversion and selectivity of products have been low compared to the Sodium phosphomolybdate salts doped with vanadium and without vanadium in its structure.

Table 5. Effect of nature of heteropoly salt catalyst on the nerol oxidation with hydrogen peroxide

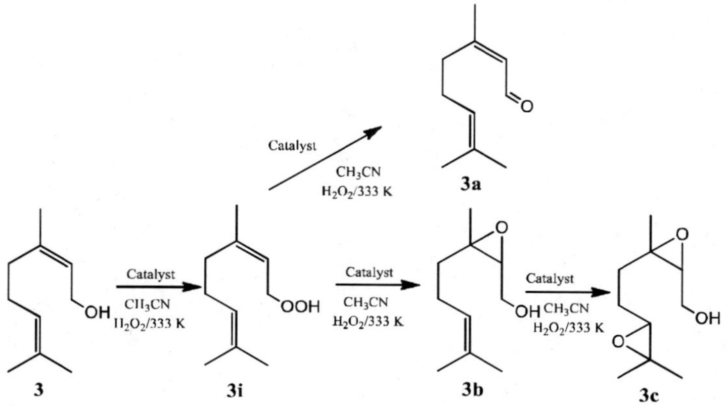

Exp.	Catalyst	Conversion (%)	Selectivity (%)			
			3a	3b	3c	3i
1	$Na_3PMo_{12}O_{40}$	75	1	68	3	29
2	$Na_4PMo_{11}VO_{40}$	96	1	96	0	3
3	$H_3PMo_{12}O_{40}$	91	7	89	0	4

[a]*Reaction conditions:* nerol: H_2O_2 (1:1); catalyst (0.66 mol %); temperature (333 K); CH_3CN (10,0 mL).

Importantly, although the Sodium phosphomolybdate salt led to the formation of diepoxide as the third product (3c), only the selectivity of the epoxide (3b) was appreciable (ca. 1% (3a), 68% (3b) and 3% (3c)). Moreover, the selectivity of nerol peroxide (3d) was much higher (ca. 28%). The heteropoly salt that contains Vanadium ($Na_4PMo_{11}VO_{40}$) presented a very good catalytic activity, which has only led to the formation of two main products (3a) and (3b), which shows its high selectivity for this catalytic system.

From the catalytic tests performed with the three Vanadium heteropolymolybdates, $Na_4PMo_{11}VO_{40}$, $Na_5PMo_{10}V_2O_{40}$, and $Na_6PMo_9V_3O_{40}$, it was possible to observe that the Sodium undecamolybdovanadate salt was the most active catalyst in the oxidation of nerol.

Neralddehyde (3a) and nerol epoxide (3b) were the main products obtained in the nerol oxidation reaction. It should be noted that epoxide nerol was the main product (ca. 96% selectivity) and aldehyde had low selectivity when the most active catalyst was used.

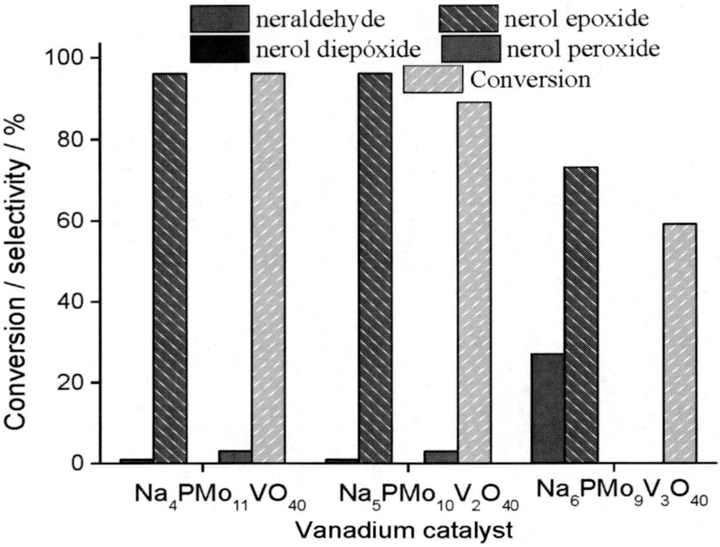

Figure 5. Effect of Vanadium content of the catalyst on the conversion and selectivity of oxidation reactions of nerol with hydrogen peroxidea. [a]Reaction conditions: nerol (2.75 mmol); H2O2 (2.75 mmol); reaction time (8 h); temperature (333K); CH3CN (10 mL).

In reactions in the presence of Sodium salts doped with two Vanadium atoms (i.e., $Na_5PMo_{10}V_2O_{40}$), only a slight decrease in conversion and selectivity was observed (Figure 5). However, when Sodium phosphomolybdate containing three Vanadium atoms was the catalyst (i.e., $Na_6PMo_9V_3O_{40}$), a lower conversion was achieved, although there was a significant increase in selectivity to neraldehyde (Figure 5).

A similar effect was observed when TiO_2-supported Vanadium-doped Cesium phosphomolybdate catalysts (i.e., $Cs_3PMo_{12-n}V_nO_{40}/TiO_2$, n = 0-3) were used in the oxidation of benzyl alcohol to benzaldehyde (RAO et al., 2015). Although a different oxidant (i.e., molecular oxygen) has been used, those authors found that while reactions catalyzed by $Cs_4PMo_{11}VO_{40}/TiO_2$ achieved the highest conversion, an increase in Vanadium content led to a decrease in conversion. Furthermore, as seen here, the aldehyde selectivity was increased. Those authors attributed this effect to the change in redox properties (ie, formation of V^{4+}/V^{5+} pairs), and the increase in proton mobility triggered by increased doping by Vanadium (RAO et al., 2015).

On the other hand, with a higher load of Vanadium (ie, V_2 or V_3 catalysts), the catalyst suffered a decline in performance, justified by the increase in the energy barrier between the HOMO and LUMO orbitals, which hinder the reducibility of these di or tri-substituted heteropolyanions (BECK et al., 2003).

The selectivity of oxidation products achieved in reactions catalyzed by Vanadium salts (i.e., mainly nerol epoxide and neraldehyde) was as follows: $Na_4PMo_{11}VO_{40}$ > $Na_5PMo_{10}V_2O_{40}$ > $Na_6PMo_9V_3O_{40}$, the same trend was observed in terms of conversion.

Table 6 shows that the selectivity of nerol peroxide was higher when greater conversions were achieved. On the other hand, the highest neraldehyde selectivity was reached in the reaction when the reaction had the lowest conversion (i.e., $Na_6PMo_9V_3O_{40}$). Again, it can be attributed to the high Vanadium content.

Table 6. Effect of Vanadium concentration on the catalyst structure for the oxidation of nerol with hydrogen peroxide

Exp.	Catalyst	Conversion (%)	Selectivity (%)			
			3a	3b	3c	3i
1	$Na_4PMo_{11}VO_{40}$	96	1	96	0	3
2	$Na_5PMo_{10}V_2O_{40}$	88	1	96	0	3
3	$Na_6PMo_9V_3O_{40}$	59	27	73	0	0

[a]Reaction conditions: nerol (2.75 mmol); H_2O_2 (2.75 mmol); time (8 h); temperature (333 K); CH_3CN (10 mL).

Doping the phosphomolybdate anion with three Vanadium atoms drastically reduces the acid strength of the catalyst, compromising its performance. One aspect that distinguishes this reaction from others is the significant formation of neraldehyde. Since these reactions (i.e., epoxidation and oxidation of alcohol to aldehyde) involve different mechanisms, we will discuss this in the next section.

It is important to highlight that the epoxidation of allyl alcohols with hydrogen peroxide in the presence of transition metal catalysts such as Tungsten, Titanium, Molybdenum or Niobium are reactions assisted by the hydroxyl group (Kumar et al., 1995; Beck et al., 2003; SonG & Barteau, 2004; Amini et al., 2014; Wang & Yamamoto, 2014; Batalha et al., 2020). This reaction competes with the oxidation of alcohol to aldehyde, and the chemoselectivity will depend on how the substrate coordinates with the metallic catalyst.

Several authors have described that, when Keggin's HPAs are the catalysts in the oxidation of alcohols or olefins with hydrogen peroxide, peroxidized intermediates are the most likely active species (Mimoun et al., 1969; Kumar et al., 1995; Patel et al., 2016; Batalha et al., 2020). When the reactions of alkenes epoxidation with hydrogen peroxide are carried out in the presence of catalysts containing Ti, hydroperoxide intermediates (ie, -OOH species) are the most common; whereas, when Vanadium, Molybdenum or Tungsten HPAs, oxo intermediates are the most favoured (ie, -OO- species) (Nojima et al., 2015; Chen et al., 2018).

Therefore, based on the literature and our experimental results, we proposed that nerol epoxidation can be described as in Scheme 2-10 (Mimoun et al., 1969; Da Silva et al., 2018; Batalha et al., 2020).

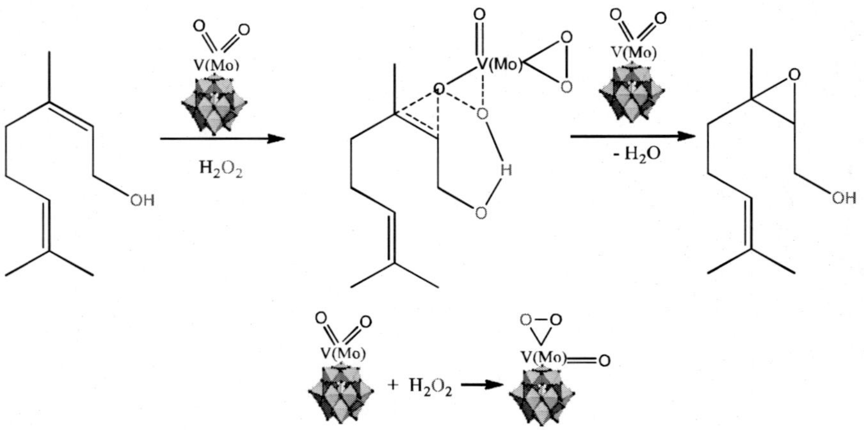

Scheme 9. Possible reaction pathway of Vanadium-doped Sodium phosphomolybdates in the epoxidation reaction of nerol by hydrogen peroxide. (adapted from Kumar et al., 1995; Da Silva et al., 2018; Batalha et al., 2020, Vilanculo and Da Silva, 2021).

Here, we assume that epoxidation is also a hydroxyl group-assisted reaction involving a phosphomolybdate doped intermediate with peroxidized Vanadium (Scheme 2-10). Although Molybdenum atoms may also be the active sites, our results showed that the formation of nerol epoxide was more favourable when Vanadium atoms were present in the Keggin anion. Romanelli et al. described that the formation of the peroxide-metal intermediate between the peroxide and the HPA catalyst (i.e., addendum atoms of Mo or W) is more favourable in the presence of Vanadium atom (Palermo et al., 2015; Palermo et al., 2016).

This beneficial effect was attributed to the energy reduction between the HOMO and LUMO orbitals, caused by the presence of a Vanadium atom, which promotes the reducibility of Keggin anions, as demonstrated in molecular orbital studies performed with vanadium-doped phosphomolybdic acids (Weber, 1994).

Table 7. Effect of $Na_4PMo_{11}VO_{40}$ catalyst concentration on the nerol oxidation reactions with hydrogen peroxide

Exp.	Catalyst load (mol %)	Conversion (%)	Selectivity (%)			
			3a	3b	3c	3i
1	0.083	74	3	44	2	51
2	0.165	84	2	69	3	26
3	0.33	89	1	83	0	16
4	0.66	96	2	96	0	2

[a]*Reaction conditions:* Nerol (2.75 mmol); H_2O_2 (2.75 mmol); tempo da reação (8 h); temperature (333 K); CH_3CN (10 mL).

It is important to highlight that even in the absence of Vanadium, the reaction occurred efficiently; therefore, it was more visible when a strong Brønsted acidity was present, which translates into better catalytic performance, as in the case of the reaction catalyzed by $H_3PMo_{12}O_{40}$ (Misono, 2001). It is noteworthy that when the catalyst concentration was increased to 0.66 mol %, a high conversion was achieved (ca. 88%) in the first hour of reaction. It is important to note that, under this reaction condition, nerol epoxide was formed as the main product (ca. 96% selectivity) and no traces of diepoxide were observed.

Therefore, although the maximum conversion has been quickly reached, it was necessary to carry out the reaction for a longer period. Furthermore, the decomposition of alkyl peroxides to oxidation product (i.e., nerol epoxide) became more efficient when a higher concentration of catalyst was present in the solution.

Conclusion

Keggin heteropolyacid salts have been demonstrated to be efficient catalysts in the reaction of epoxidation of terpene alcohols. The doping of heteropolyanion with transition metal cations enhance the performance of these catalysts, increasing also the reaction selectivity toward epoxides. Particularly, the selectivity toward epoxide is much higher when allylic alcohols are the substrates. Lacunar heteropolyacid salts are also an attractive option to be used as catalysts in epoxidation reactions with hydrogen peroxide.

Lacunar Sodium phosphotungstate and Cesium silicotungstate converted linalool, nerol, and geraniol to epoxides. On the other hand, the doping of phosphomolybdic acid with Vanadium cations resulted in very active catalysts, which were successfully used in the nerol epoxidation. The use of an environmentally benign oxidant (i.e., hydrogen peroxide), and easily synthesizable solid catalysts (heteropolyacid salts) are the positive aspects of this selective route to obtain terpene alcohols epoxide.

References

Ali, M.; Hazmi, A. L.; Nadeem S. S.; Sheikh, Bataille, C. J. R.; Al-Hadedi, A A. M.; Watkin, S. V.; Luker, T. J.; Camp, N. P.; Brown. R. C. D. *Organic Lett*ers, 16, 5104-5107, 2014.

Amini, M.; Haghdoost, M. M.; Bagherzadeh, M. *Coordination Chemistry Reviews*, 268, 83-100, 2014.

Amitouche, D.; Haouas, M.; Mazari, T.; Mouanni, S.; Canioni, R.; Rabia, C.; Cadot, E.; Marchal-Roch, C. *Applied Catalysis A*., 561, 104–116, 2018.

Astruc, D. Part IV: Catalysis, in: *Organometallic chemistry and catalysis*. Berlin: Springer. V. 291, 2007.

Balula, S. S.; Santos, I. C. M. S.; Silva, L. C.; Carvalho, A. P.; Pires, J.; Freire, C.; Cavaleiro, J. A. S.; Castro, B.; Cavaleiro. A. M. V. *Catalysis Today*, 203, 95-102, 2013.

Barteaua K. P.; Lyonsa J. E.; Songb I. K.; Barteau, M. A. *Topics in Catalysis,* 41(1–4), 55-62, 2006.

Batalha D. C. B.; Marins, N. H.; Silva, R. M.; Carreño, N. L. V.; Fajardo, H. V.; Da Silva M. J. *Molecular Catalysis,* 489, 110941-110952, 2020.

Beck, C.; Mallat, T.; Baiker, A. *Catalysis Letters,* 88, 203-209, 2003.

Bernardo-Gusmão, K.; Pergher, S. B.; Santos, E. N. *Química Nova*, 40(6), 650-655, 2017.

Boivin, T. L. B. *Tetrahedron*, 43, 3309–3362, 1987.

Bond, G. C.; Frodsham, S. J.; Jubb, P.; Kozhevnikova, E. F.; Kozhevnikov, I. V. *Journal of Catalysis,* 293, 158-164, 2012.

Borras-Almenar J. J.; Coronado, E.; Muller, A.; Pope, M. T. *Polyoxometalate Molecular Science,* Kluwer Academic Publisher, Dordrecht, 2003.

Bryliakov, K. P. *Chemical Reviews,* 117(17), 11406-11459, 2017.

Casuscelli, S. G.; Crivello, M. E.; Perez, C. F; Ghione, G.; Herrero, E. R.; Pizzio, L. R.; Vásquez, P. Z.; Cáceres, C. V.; Blanco, M. N. *Applied catalysis A: General,* 274(1-2),115-122, 2004.

Cavani, F. *Catalysis Today,* 41(1–3), 73-86, 1998.

Chen, Q.; Shena C.; Hea, L. *Acta Crystollagraphica,* C74, 1182-1201, 2018.

Chorkhendorff, I. Niemantsverdriet. J. W. *Concepts of modern catalysis and kinetics*, 2003.

Clayden, J, Greeves, N, Warren, S, Wothers. P. *Organic Chemistry.* Oxford, 2000.

Coronel, N. C.; Da Silva, M. J. *Journal of Cluster Science*, 29, 195-205, 2018.

Da Silva, M. J. *Current Organic Chemistry*, 20, 1263-1283, 2016.

Da Silva, M. J.; Andrade, P. H. S.; Ferreira, S. O.; Vilanculo, C. B.; Oliveira, C. M. *Catalysis Letters*, 148, 2516, 2018.

Da Silva, M. J.; Carari, D. M.; Da Silva, A. M. *RSC Advances,* 5(14), 10529-10536, 2015.

Da Silva, M. J.; Gonçalves, J. A.; Alves, R. B.; Howarth, O. W, Gusevskaya, E. V. *Journal of Organometallic Chemistry*, 689, 302-308, 2004.

Da Silva, M. J.; Gonçalves, J. A.; Alves, R. B.; Howarth, O. W.; Gusevskaya, E. V. *Journal of Molecular Catalysis. A: Chem.*, 176, 23, 2001.

Da Silva, M. J., Liberto, N. A., Leles, L. C. A. *Journal of Molecular Catalysis A: Chemical*, *422*, 69-83, 2016.

Da Silva; M.; De Oliveira, C. M. *Current Catalysis, 7,* 26-34, 2018.

De Meireles; A. L. P; Da Silva Rocha, K. A; Kozhevnikov, I. V; Gusevskaya, E.V. *Applied Catalysis A: General*, 409–410, 82-86, 2011.

De Oliveira, A. A.; Da Silva, M. L.; Da Silva, M. J. *Catalysis Letters*, 130, 424-431, 2009.

De Vos, D. E.; Sels, B. F.; Jacobs, P. A. *Advanced Synthesis and Catalysis,* 345, 457-473, 2003.

Duprez, D. & Cavani, F. (Eds.), *Handbook of Advanced Methods and Processes in Oxidation Catalysis,* Imperial College Press, London, 2014.

Dusi, M.; Mallat, T.; Baiker, A. *Catalysis Reviews,* 42, 213-278, 2000.

Farsani, M. R., Yadollahi, B. *Journal of Molecular Catalysis A: Chemical, 392*, 8–15, 2014.

Gallezot, P. *Chemical Society Reviews*, 41(4), 1538–1558, 2012.

Guo Z.; Liu, B.; Zhang, Q.; Deng, W.; Yang, Y. *Chemical Society Reviews,* 43(10), 3480-3524, 2014.

Haider, M. H.; Dummer, N. F.; Zhang, D.; Miedziak, P.; Davies, T. E.; Taylor, S. H.; Willock, D. J.; Knight, D. W.; Chadwick, D.; Hutchings, G. J. *Journal of Catalysis*, 286, 206-213, 2012.

Jing, F.; Katryniok, B.; Dumeignil, F.; Bordes-Richard, E.; Paul. S. *Journal of Catalysis,* 309, 121-135, 2014.

Kamata, K.; Kotani, M.; Yamaguchi, K.; Hikichi, S.; Mizuno, N. *Chemistry A European Journal,* 13(2), 639–648, 2007.

Karcz, R.; Niemiec; P.; Pamin; K.; Kryściak-Czerwenka; J.; Napruszewska; B. D.; Michalik-Zym, A; Witko; M.; Tokarz-Sobieraj; R.; Serwicka, E. M. *Applied Catalysis A: General*, 542,5, 317-326, 2017.

Kourasi, M; Wills, R. G. A.; Shah, A. A.; Walsh, F. C. *Electrochimica acta*, 127, 454-566, 2014.

Kozhevnikov, I. V. *Journal of Molecular Catalysis A.*, 117, 151-158, 1997.

Kumar, R.; Pais, G. C. G.; Pandey, B.; Kumar, P. *Journal of the Chemical Society, Chemical Communications,* 13, 1315-1316, 1995.

Lee, J. K.; Melsheimer, J.; Berndta, S.; Mestl, G.; Schlögl, R.; Köhle K. *Applied Catalysis A.*, 214, 125-148, 2001.

Li, L.; Liu, B.; Wu, Z.; Yuan, X.; Luo, H. *Chemical Engineering Journal,* 280, 670–676, 2015.

Lopez, X.; Carbo, J. J.; Bo, C.; Poblet, J. M. *Chemical Society Reviews*, 41, 7537-7571, 2012.

Lv, H.; Song, J.; Zhu, H.; Geletii, Y. V.; Bacsa, J.; Zhao, C.; Lian, T.; Musaev, D. G.; Hill, C. L. *Journal of Catalysis*, 307, 48-54, 2013.

Maksimchuk, N. V.; Melgunov, M. S.; Mrowiec-Białoń, J, Jarzębski, A. B.; Kholdeeva, O. A. *Journal of Catalysis,* 235, 175-183. 2005.

Marin-Astorga, N.; Martinez, J. J.; Borda, G.; Cubillos, J.; Suarez, D. N.; Rojas, H. *Topics in Catalysis*, 55, 620-624, 2012.

Mekki-Berrada, A.; Bennici, S.; Gillet, J. P.; Coutourier, J. L.; Dubois, J. L.; Auroux, A. *Journal of Catalysis,* 306, 30-37, 2013.

Mimoun, H.; De Roch, I.S.; Sajus, L. *Bull. Soc. Chim., Fr.,* 1481, 1969.

Misono, M. *Chemical. Communications*, 1141-1152, 2001.

Mizuno, N. & Kamata, K. *Coordination Chemistry Reviews,* 255, 2358-2370, 2011.

Narkhede, N.; Singh, S.; Patel, A. *Green Chemistry,* 17, 89-107, 2015.

Nasir, N. M.; Ermanis, K.; Clarke, P. A. *Organic and Biomolecular Chemistry,* 12, 3323–3335, 2014.

Nicolaou, K. C.; Prasad, C. V. C.; Somers, P. K.; Hwang, C. K. *Journal of American Chemical Society,* 111, 5330–5334, 1989.

Nojima, S.; Kamata K.; Suzuki, K.; Yamaguchi, K.; Mizuno M. *Chem Cat Chem,* 7, 1097-1104, 2015.

Norskov, J. K.; Studt, F.; Pedersen, F. A.; Bligaard, T. *Fundamental concepts in heterogeneous catalysis*. Willey, 2014.

Omwoma, S.; Gore, C. T.; Ji, Y.; Hu, C.; Song, Y. F. *Coordination Chemistry Reviews,* 286, 17–29, 2015.

Oro, L. A.; Sola, E. *Fundamentos y Aplicaciones de la Catálisis Homogénea*, 1ª ed; Zaragoza (2000), p. 208. [*Fundamentals and Applications of Homogeneous Catalysis*]

Palermo, V.; Sathicq, Á. G.; Constantieux, T.; Rodríguez, J.; Vázquez, P. G.; Romanelli, G. P. *Catalysis Letters,* 146, 1634-1647, 2016.

Palermo, V.; Sathicq, Á.; Constantieux, T.; Rodríguez, J.; Vázquez, P.; Romanelli, G. *Catalysis Letters,* 145, 1022-1032, 2015.

Patel, A.; Narkhede, N.; Singh, S.; Pathan, S. *Catalysis Reviews,* 58(3), 337-370, 2016.

Pope, M. T. *Inorganic Chemistry Concepts Vol. 8: Heteropoly and Isopoly Oxometalates*, Springer, Berlin, 1983.

Rao, P. S. N.; Parameswarama, G.; Rao, A. V. P.; Lingaiah, N. *Journal of Molecular Catalysis A.,* 399, 62-70, 2015.

Sadri, F.; Ramazani, A.; Massoudi, A.; Khoobi, M.; Tarasi, R.; Shafiee, A.; Azizkhani, V.; Dolatyari, L.; Joo, S. W. *Green Chemistry Letters and Reviews*, 7(3), 257-264, 2014.

Serra, S. &. De Simeis. D. *Catalysts,* 8, 362-378, 2018.

Somma, F.; Strukul, G. *Journal of Catalysis,* 227, 344-351, 2004.

Song, I. K. & Barteau, M. A. *Journal of Molecular Catalysis A*., 212, 229-236, 2004.

Song, X., Zhu, W., Li, K., Wang, J., Niu, H., Gao, H., Jia, M. *Catalysis Today, 259*, 59–65, 2016.

Tsolakis, N.; Bam, W.; Srai, J. S.; Kumar, M. *Journal of cleaner Production*, 222, 802-822, 2019.

Viana, L. A. S.; Da Silva G. R. N.; Da Silva, M. J. *Catalysis Letters,* 148, 374-382, 2018.

Vilanculo C. B. & Da Silva. M. J. *New Journal of Chemistry,* 44, 2813-2820, 2020.

Vilanculo, C. B.; Da Silva, M. J.; Teixeira M. G.; Villarreal, J. A. *RSC Advances,* 10, 7691-7697, 2020.

Vilanculo, C. B.; Da Silva, M. J.; Rodrigues, A. A.; Ferreira, S. O.; Da Silva, R. C. *RSC. Advances*, 11, 24072–24085, 2021.

Wang C. & Yamamoto, H. *Journal of American Chemical Society,* 136, 1222-1225, 2014.

Weber, R. S. *Journal of Physical Chemistry,* 98, 2999-3005, 1994.

Zhang, A. P.; Zhang, Y.; Wang, P. G.; Li, J.; Lv, Y.; Gao, S. A. *Research on Chemical Intermediates*, 37, 975-984, 2011.

Zheng, H.; Yan; Z.; Chu, S.; *J. Chen; Chemical Engineering and Processing - Process Intensification*, 134, 1-8, 2018.

Chapter 4

Terpenoids: The Phyto-Active Components as Sources of Insecticide

Bhabesh Deka[1,*], Archana Borah[2], Uma Dutta[3] and Azariah Babu[1]

[1]Department of Entomology, North Bengal Regional Research and Development Centre, Nagrakata, West Bengal, India
[2]Department of Zoology, B. Borooah College, Guwahati, Assam, India
[3]Department of Zoology, Cotton University, Guwahati, Assam, India

Abstract

Natural products derived from plant sources have a wide range of biological applications. They are eco-friendly and mitigate or prevent many health ailments and environmental changes due to synthetic products. The long and persistent use of synthetic pesticides imposes a serious threat to non-target organisms and contaminates soil and groundwater. The plant-based products are bio-degradable in nature and could be considered as an alternate means of pesticides. Terpenoids are the secondary metabolites of plant derivatives, mainly found in essential oils derived from higher plant sources. They act on insects in various ways, mainly as chemical defenses, insect repellents, attractants, insecticides, and anti-feedents. Some are reported to influence the growth and development of some insects, which further leads to mortality and reproduction. Being volatile and lipophilic in nature, terpenoids could be

* Corresponding Author's E-mail: bhabesh.deka@gmail.com.

In: Terpenes and Terpenoids
Editor: Charles A. Davies
ISBN: 978-1-68507-559-0

used as fumigants which easily penetrate the insect body and might interfere with physiological, biochemical, and behavioural functions. Some terpenoids are phytotoxic to plants. This toxicity could be reduced by chemical modification of the parent compound. However, some of the derived compounds after chemical modification show enhanced pesiticidal activity. The understanding of the chemical nature of the compound and the influence of their physicochemical parameters are essential to predict the toxicological aspects of terpenoids and their insecticidal property.

Keywords: Terpenoid compounds, Natural products, Insecticide, Pest management

Abbreviations

FPP	Farnesyl Pyrophosphate
GABA	Gamma-Aminobutyric Acid
GFPP	Geranyl Farnesyl Diphosphate
GGPP	Geranyl-Geranyl Pyrophosphate
GPP	Geranyl Diphosphate
JA	Jasmonic Acid
MEP	Methylerythritol Phosphate
MRL	Maximum Residue Limits
nAChR	Acetylcholine Esterase Nicotinic Acetylcholine Receptors
SA	Salicylic Acid
THCV	Tetrahydrocannabinol Volatiles

Introduction

In different temperate zones, the insect community interacts with a range of plants. Negative interactions with insects have resulted in the development of several complex chemical defence systems (Fürstenberg-Hägg et al., 2013). Insect diversity is determined by plant diversity and evolution. By 2050, the food supply will have to be tripled to meet demand, according to the forecast (Parry 2010). Reduced agricultural production is a result of arthropod pest damage (Dubey et al., 2014). The annual yield loss in the world due to pest attack surpasses 15%. (Mendesil et al., 2014). Pesticide use has resulted in several negative consequences, including unintended consequences for humans and beneficial insects and pollinators. Herbivory triggers a range of plant defense systems and the induction of a defensive response will form an important part of pest insect control. As a result of insect bites, Ca++ fluxes and phosphorylation cascades are generated (Fürstenberg-Hägg et al., 2013). Plants respond to herbivores by producing a variety of defence chemicals such as phytoanticipins and phytoalexins, which are inactive metabolites, while phytoanticipins and phytoalexins induce an active response. Insects are repelled or intoxicated by bioactive compounds, which affect their digestion. Anti-herbivore defences based on carbon or nitrogen act as a deterrent, growth inhibitor, repellent, or cause direct death of the insects. In response to the salicylic acid (SA), ethylene, and jasmonic acid (JA) pathways, defensive proteins against herbivorous insects are synthesized with the purpose of detecting and preventing the spread of biological agents that could cause serious harm to crops. Terpenoids, also known as isoprenoids, are a diverse collection of natural compounds generated through a common biochemical pathway intermediate and created by condensation of five-carbon isoprenoid units into mevalonate as the parent molecule (Duke, 1991). Higher plants, mosses, liverworts, algae, and lichens, as well as a few insects and microorganisms, are their major sources. They are classified into several classes depending upon the parent compound they are derived from. The terpenoids derived from GPP10, FPP15,GGPP20, GFPP25, Squalene30, Phytoene40 and GGPP103-104 are named as monoterpenoids, sesquiterpenoids, diterpenoids, sesterpenoids, triterpenoids, carotenoids and rubber respectively (Banthorpe, 1991). These groups of compounds are mainly targeted for pest control, especially insect pests.

Chemical Nature and Biological Property of Terpenoids

Terpenoids, which are naturally occurring hydrocarbons, are produced by a broad variety of plants and animals. Since the discovery of five-carbon (isoprene) molecules, more than 55,000 compounds have been discovered. Terpenes include hemiterpenes (C5), monoterpenes (C10), sesquiterpenes (C15), diterpenes (C20), sesterterpenes (C25), triterpenes (C30), and polyterpenes (>C30) (Brahmkshatriya and Brahmkshatriya 2013). Terpenoids are produced by two distinct pathways: the MVA pathway and the MEP pathway (Brahmkshatriya and Brahmkshatriya 2013). Terpenoids have antibacterial, antifungal, antiviral, antihyperglycemic, anti-inflammatory, antioxidant, antiparasitic, immunomodulatory, and skin permeability properties (Manchali et al., 2011).

Role of Terpenoids in Insect-Plant Interaction

Plants generate a large number of secondary metabolites that protect the plants against their enemies (herbivores). Terpenoids are plant-derived secondary metabolites that influence plant-insect interactions, which are believed to be advantageous for plants.

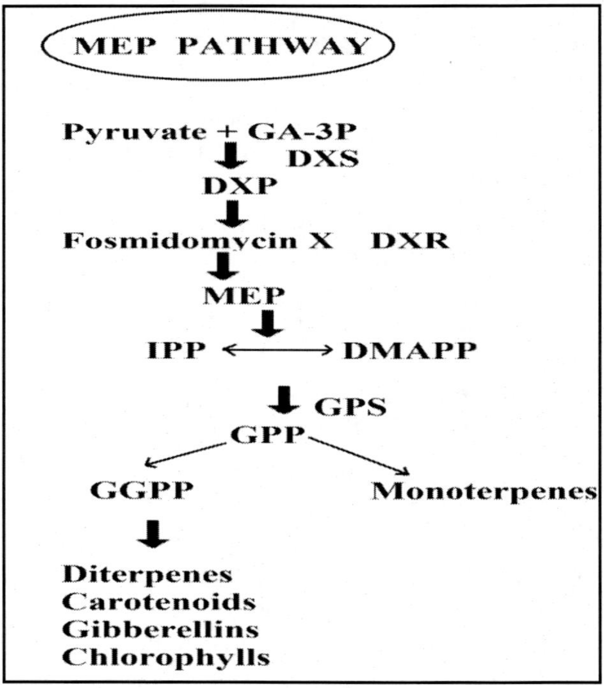

Figure 1. The methylerythritol phosphate (MEP) pathways in biosynthesis of terpenoid.

It was previously believed that such interactions are mediated via the emission of volatile organic molecules. Terpenoids are a diverse category of volatile oils that plants produce in significant quantities (Dudareva and Pichersky 2006). Terpenoids' volatile components are dominated by the hemeterpene isoprene (C5), monoterpenes (C10), sesquiterpenes (C15), and their derivatives, such as homoterpenes (C11 and C16). Phytochemical characteristics of terpenoids such as volatility, reactivity, toxicity, and fragrance contribute to the terpenoid's protective qualities against abiotic and biotic causes (Holopainen 2004).

It is thought that the properties of volatile organic compound mixtures are determined by the identity and quantity of individual components, and the attacking herbivores are sensitive to certain volatile organic compound blends (Boncan et al., 2020). It has been shown in many studies that herbivorous animals alter the emission of volatile mixtures in a wide variety of plant species (Dicke and Van Loon 2000; Dicke and Van Poecke 2002; D'Alessandro and Turlings 2006). Herbivore-induced alterations in volatile blends may be quantitative (Mumm et al., 2003; Bukovinszky et al., 2005). However, herbivores can also drive de novo synthesis of chemicals in a wide variety of plant species, resulting in qualitative changes in the composition of the emitted blend (Turlings et al., 1998; Dicke et al., 1999; Leitner et al., 2005).

Plants' defence mechanisms rely on terpenes, a naturally occurring hydrocarbon. This class includes all plants, with roughly 22,000 unique chemicals (some examples are shown in Table 1 and 2). Isoprene is the backbone of this class of terpenes. Most terpenes are produced via the MEP route (Figure 1). Besides, terpenes add to the aroma of plants. The simplest forms of terpenoids are those associated with photosynthesis in plants. To combat insect pests, most coniferous trees utilize terpenoids and polyphenols in the resin canals of the xylem (Boncan et al., 2020). Glycoside precursors are utilized to generate steroids and sterols. Additional elements influence terpene synthesis progression. Terpenes help plants ward off insect herbivores by providing signals to predators, which in turn control the herbivores either by predating or parasitism channeled through a specific volatile combination (Bond et al., 2004).

Monoterpenoids and Sesquiterpenoids

Volatile monoterpenoids and sesquiterpenoids (Figure 2) keep plants free from pests. Woolly adelgid resistance is increased by hemlock resin volatiles (THCV). This might occur due to accumulation of Benzyl alcohol and methyl salicylate (Keeling and Bohlmann 2006). Conifers create more monoterpenes and sesquiterpenes in the presence of pests or diseases. Menthol and menthone are both made from the epidermal glandular trichomes of mint plants (Mentha spp.). Pyrethrins can poison some insects (also known as "Chrysanthemum pyrethrins") (Pezet and Elkinton 2014). Some synthetic pyrethrum alternatives are known as pyrethrum substitutes (cypermethrin and permethrin). Many spices, sauces, condiments, and perfumes use monoterpenoids (Keeling and Bohlmann 2006). However, they are not toxic to humans. Spearmint, basil, oregano, rosemary, sage, and savoury all kill insects. Orange oil derived terpenes and neem oil were reported to be efficient at controlling greenhouse aphids (Pezet and Elkinton 2014).

Diterpenoids

The non-mevalonic acid pathway is responsible for the production of diterpenes (Figure 2) and carotenoids. Gossypol contains isoprene units, which can be found in latex and resin, as well as in other materials (Pham et al., 2014). The poisonous units have a repellent effect on insects and Rhododendron poisoning is caused by diterpenes. All plants include rhododendron and romedotoxin, which are reported to be toxic to humans. Nectar contains romedotoxins, which inhibit the activity of Na^+ channels (Singh and Sharma, 2015). Spinach has a negative impact on both larval development and death (*Spinacia oleracea*). Insects are repelled by the strong citrus odours, which contain limonoids. The limonene terpene, which is abundant in citrus fruits, on the other hand, is known to attract insects (Mithöfer and Boland 2012).

Table 1. Terpenoids, their source plants and activities in insect-plant interactions (Senthil-Nathan 2013; TlakGajger and Dar 2021)

Chemical	Source Plant	Affected Pest Insect	Effect
α-Pinene 3-Carene	*Pinus silvestris*	*Blastophagus piniperda*	Repellent
α-Pinene 3-Carene	*Pinus taeda*	*Dendroctonus frontalis*	Attractant
α-Pinene 3-Carene	*Pinus ponderosa*	*Dendroctonus brevicomis, Heliothis* spp.	Attractant
Gossypol	*Gossypium hirsutum*	*Epicauta* spp.	Feeding deterrent
		Anthonomus grandis	Feeding stimulant
		Tetranychus urticae	Feeding deterrent
Cucurbitacin	Cucurbitaceae	*Acalymma* spp. *Diabrotica* spp.	Feeding excitant

Table 2. Different plant species with common volatile terpenes released in response to insects' attack ('+': Active response; '–': Non-response)

Plant	**Linalool**	**(E)-4,8-Dimethyl-1,3,7-Nonatriene**	**(E,E)-a-Farnesene**	**(E)-b Ocimene**	**(E)-b-Farnesene**	**(E,E)-4,8,12-Trimethyl 1, 3, 7, 11-Tridecatetraene**	**Reference**
Cotton	+	+	+	+	+	+	Jawale, 2014
Corn	+	+	+	+	+	+	Takabayashi and Dicke, 1994
Tobacco	+	-	+	+	+	-	Llandres et al., 2018
Potato	+	-	+	+	+	-	Pandian et al., 2020
Apple	-	+	+	+	-	+	Mbata, and Payton 2013
Cucumber	-	+	+	+	-	+	Jawale, 2014
Lima bean	+	+	-	+	+	+	Jawale, 2014

The neem trees (*Azadirachta indica*) and lemon grass (*Cymbopogon citratus*) contains potent limonoids, Azadirachtin and Citronella, which have been derived from them. Insect repellents such as citronella have limonoids as their primary active component. Citronella is non-toxic to humans and completely biodegradable in the environment. *A. indica* based insect repellents and food-deterrent compounds are available on the market. Most of these products contain a-Phellandrene, camphene, chrysanthenyl acetate, borneol, camphor, and a-Phellandrene (Wink 2009).

Saponins

Saponins, also known as glycosylated triterpenoids (Figure 2), are found in the membranes of plant cells. Its detergent-like activity breaks down cell membranes, kills nuisance insects, and induces cell death in a variety of bacteria and viruses (Kazemi 2014). Saponins have an interaction with cholesterol, which prevents the production of steroid hormones from occurring. Saponins are found in all legumes, and many of them are insecticidal in nature (repellent/deterrent). The most frequently reported symptoms include limited reproduction, mortality, developmental delays, and weight loss of targeted insects. Seed pods containing saponins have been introduced in order to discourage birds from consuming the seeds of desert shrubs (Chaieb 2010). Mosquito repellent saponins are found in the fruits, leaves, and flowers of the plant *Sapindus mukorossi*, *Cestrum nocturnm*, *C. diurnum* and *Asclepias curassavica* and they can be used to get rid of mosquitoes (De Geyter et al., 2007).

Insecticidal Activities of Terpenoids and Their Derivatives

By the late 1970s, research on the biological action of terpenoids had advanced significantly. It was in the late 1990s, after the discovery of artemisinin, that the study of terpenoids gained prominence, with emphasis on the separation and purification of active components and chemical structure (Yang et al., 2020).

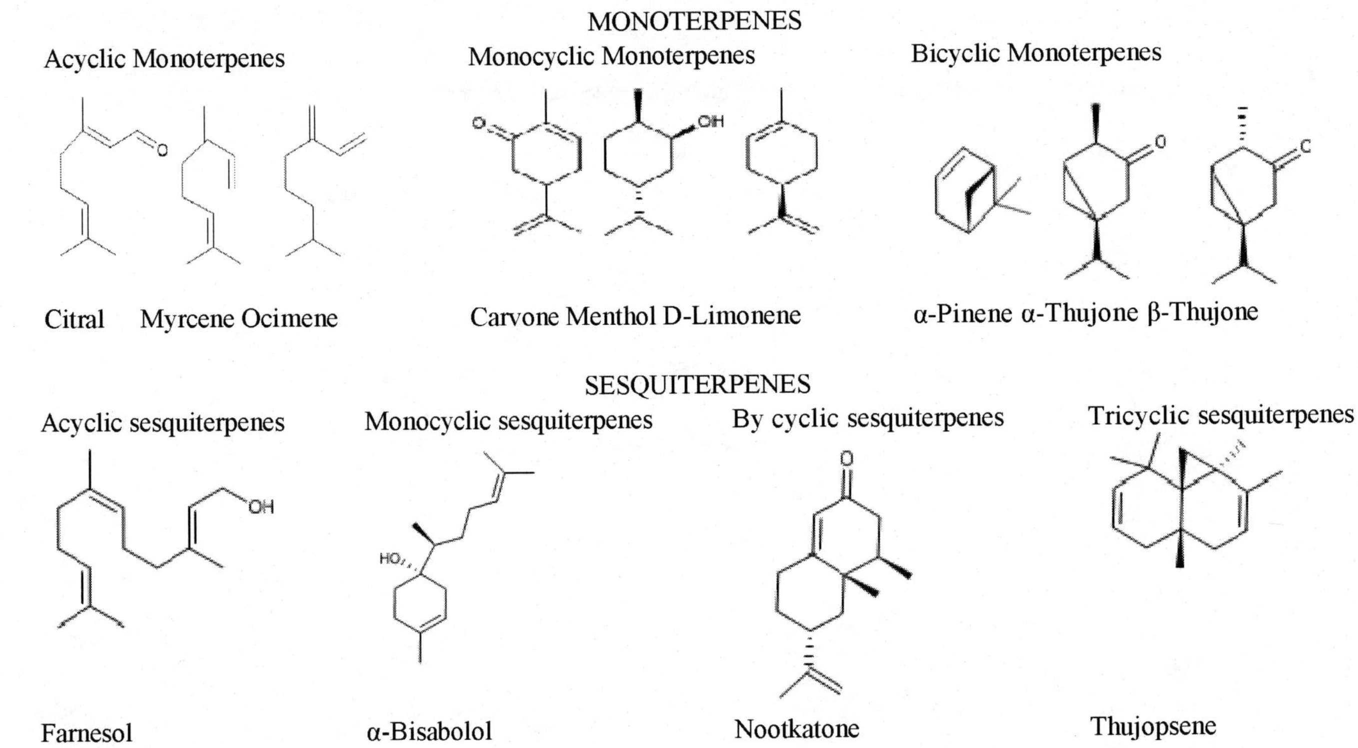

Figure 2. (Continued)

DITERPENES

Linear Diterpenes	By cyclic Diterpenes	Tricyclic Diterpenes	Tetracyclic Diterpenes
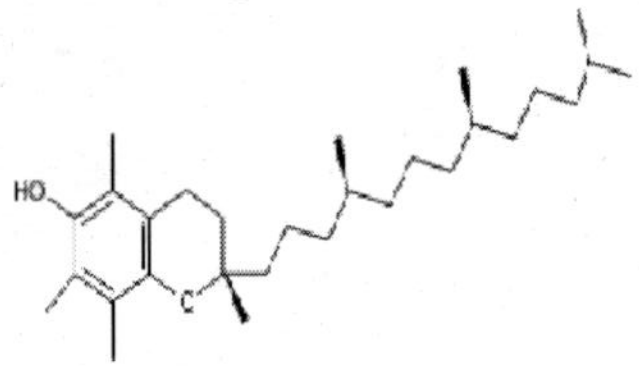	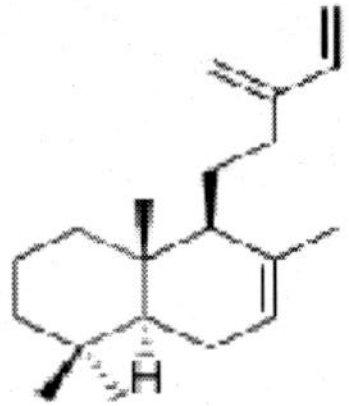	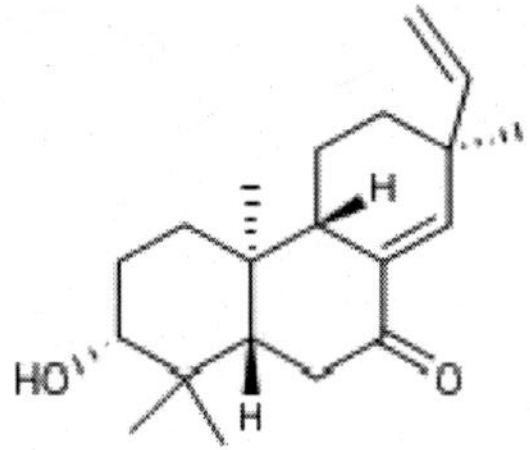	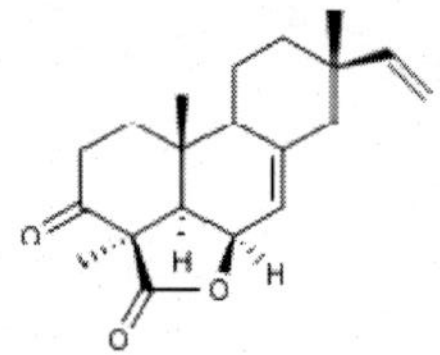
α-Tocopherol	Labda-7,13(16),14-triene	Oryzalexin A	Momilactone A

TRITERPENES

Linear Triterpenes	Bicyclic Triterpenes	Tricyclic Triterpenes	Tetracyclic Triterpenes	Pentacyclic Triterpenes
	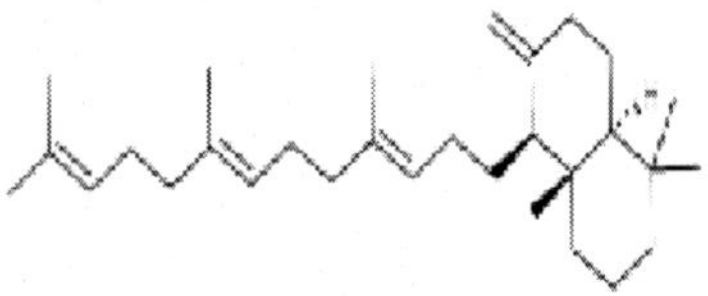	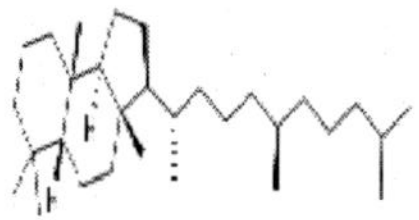	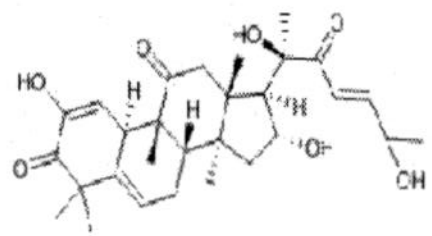	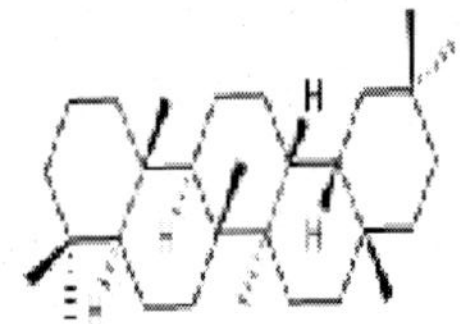
Squalene	α-polypodatetraene	Cucurbitacin	Malabaricane	Oleanane

Figure 2. Structural forms of volatile terpenes: Monoterpenes; Sesquiterpenes, Diterpenes and Triterpenes. (Structures were drawn and analyzed with ChemDraw software, version 20.0.0.42).

Cinnamon and clove essential oils are reported to be poisonous and repellant to *Stophilus granaries*, a significant grain pest. Eugenol has the highest level of contact toxicity (Plata-Rueda et al., 2018). Cucurbitacins, which are linked to cucumber's bitterness, are poisonous to a most of the organisms and are found in both wild and farmed cucurbitaceae. Cucurbitacin C, isolated from *Cucumis sativus*, was shown to be resistant to Tetranychusuricae, a spider mite (Balkema-Boomstra, 2003). Geraniol was shown to be a more effective repellent against four distinct mosquito species compared to citronella or linalool (Muller et al., 2009).

When naturally occurring monoterpenoid toxicity was evaluated against several insects, citronellic acid and thymol were found to be more lethal topically against the house fly, whereas citronellol and thujone were found to be more effective against the western corn rootworm. Carvomenthenol and terpien-4-ol were shown to be very efficient against two species of spotted spider mite. Perillaldehyde was shown to have strong larvicidal efficacy against western corn rootworm larvae (Lee et al., 1997). Among the plant-derived methyl ketones: 2-heptanone, 2-octanone, 2-nonanone, and 2-undecanone, undecanone exhibited the most potent fumigant action against houseflies (Deguenon et al., 2019).

Numerous research on diterpenoids derived from plants has also shown that, they possess insecticidal properties. Diterpenoids' insecticidal activity is mediated primarily through their antifeedant and repellent properties. Cassane diterpenoids derived from aerial parts of *Caesalpinia pulcherrima* has shown remarkable antifeedant action against *Mythimna separate*. Whereas, Pulcherrimin C and 12-dimethyl neocaesalpin F were found to be effective antifeedants against *Plutella xylostella* (Li et al., 2020).

The antifeedant activity of ent-kaurane diterpenoids isolated from Turkish Sideritis species was evaluated against Lepidoptera final stadium larvae (Aslam et al., 2006). Among them, sideroxol, which was isolated from the majority of the Sideritis species, demonstrated significant antifeedant activity against *Spodoptera frugiperda*, and another ent-kauranefoliol, which is also found in Sideritis species, was reported to be a potent phagostimulant for *Spodoptera littoralis* (Bondi et al., 2002). Ent-kaurene diterpenoids, 7-epicandicandiol, 18-acetylsideroxol, and an acetone extract of *Sideritis trojana* were poisonous to *Acanthoscelides obtectus* and *Sitophilus granarius*, while 7-epicandicandiol diacetate was hazardous exclusively to *Ephestia kuehniella* (Aslam et al., 2006). Linearol acetone extracts of *S.lycia* were shown to be efficacious against *Tetranychus urticae*. Diterpenoid clerodin present in *Clerodendron infortunatum* possessed significant antifeedant activity against the cotton bollworm, *Helicoverpa armigera* (Abbaszadeh et al., 2014).

Table 3. Insecticidal activities of terpenes

Name of Terpene	Plant	Activity	Reference
β-pinene	*Citrus paradisi, Poncirus trifoliata*	Attracts entomopathogenic nematodes, e.g., *Steinernema diaprepesi*	Ali et al., 2011
β-ocimene	Tomato and tobacco	Defense against pests, e.g., *Macrosiphum euphorbiae* (potato aphid)	Cascone et al., 2015
1,8-cineole	Brassica Tropical orchids	Attracts egg-laying parasitoids to the caterpillarsof feeding herbivores. Furthermore, attracts and reward pollinators	Byers et al., 2014
Decanal	Zea mays	Enhanced resistance against *Ostrinia nubilalis*	Solé et al., 2010
β-trans-ocimene, (+)-R-limonene	Lavender	Deters pests, e.g., aphids	Wu et al., 2018
Eucalyptol	Oak	Attracts cockchafer larva	Weissteiner et al., 2012
S-methyl methionine	Lecythidaceae	Deters oviposition sites-seeking beetle	Bian et al., 2017
Rhizathalene A	Arabidopsis	Resistance of the roots to the herbivore dark-winged fungus gnat (*Bradysia* spp)	Vaughan et al., 2013
Eugenol, caryophyllene oxide, α-Pinene, α- humulene, and α -phellandrene	Cinnamon and clove	Secret toxic terpenes to deter the adult pest of *Sitophilus granaries* (grain weevil)	Plata-Rueda et al., 2018
β-costic acid	Zea mays	Inhibits the growth of *Diabrotica balteata* (cucumber beetle)	Ding et al., 2017

Among sesquiterpenoid lactones, sesquiterpene lactones from the fruits of *Carpesium abrotanoides* possessed antifeedant activity against 3rd instar *Plutella xylostella* larvae as well as stomach contact toxicity against 4th instar *Bradysia odoriphaga* larvae (Wu et al., 2016). Carpariolides A-D, novel structurally isomeric sesquiterpenoids from Carpraria biflora, was found to have insecticidal action against adults of the sweet potato weevil *Cylas formicarius elegantulus* (Collins et al., 2002).

Titerpenoids have also been shown to have insecticidal effects, as highlighted in several studies. Two novel triterpenoids, 22, 23-dihydronimocinol and desfurano-6-hydroxyazadiradione, were extracted from a methanolic extract of fresh leaves of *Azadirachta indica* and were shown to exhibit insecticidal activity against *Anopheles stephensi*, killing fourth instar larvae (Siddiqui et al., 2002). The 2α,2β, 21β, 23, 28-penta hydroxyl 12-oleanene isolated from the roots of *Laportea crenulata* Gaud has both pesticidal and repellent activity against *Tribolium castaneum*, a significant stored grain pest (Khan et al., 2014). Titerpenoids, namely 2E-3β-(3, 4-dihydroxycinnamoyl)-erythrodiol, 2Z-3 β-(3, 4-dihydroxycin-namoyl)-erythrodiol and 2E-3β (3, 4-dihydroxyphenyl) - 2- propenoate is an insecticide derived from the volatile oil of *Tetraena mongolica* stems. It has antifeedant and growth inhibitory properties (Wu et al., 2020). Table 3 depicts the role of terpenes as a potent agent of defense against insects.

Mode of Action of Insecticidal Activity

The precise mechanism of action of terpenoids' insecticidal effect is yet to be understood. Numerous studies have indicated that, insecticidal activity demonstrated as a repellant reduces food intake, inhibits development, has membrane permeabilising properties, and induces apoptosis, among other effects. Terpenoid has been reported to show a dose-dependent antifeedant action (Omar et al., 2007). Antifeedant chemicals exert their effect on the pest's sensory cells. These neurons' receptors either inhibit insect feeding or induce the stop or slowing of future feeding. Another mechanism of action for certain antifeedants is their apparent capacity to inhibit the operation of herbivores' feeding stimulant receptors or to bind directly to the herbivore's natural feeding signals, such as sugar and amino acids (Purrington, 2003). Certain chemicals exert their impact through contact chemoreception and internal feedback mechanisms that promote enzyme synthesis, cell proliferation, and motility. Azadiractin is a good example of this mode of action (Mordue et al., 2005).

Numerous researches on monoterpenoid esters have reported its dose-dependent repellency (Born et al., 2012; Nesterkina et al., 2018). The precise mechanism by which repellants work is not completely known. Monoterpenoids have been proposed to act on the insect nervous system through GABA-gated chloride channels, octopamine receptors, tyramine receptors, acetylcholine esterase nicotinic acetylcholine receptors (nAChR), and sodium channels, among others (Coats et al., 1991; Kostyukovsky et al., 2002). Certain monocyclic terpenoids serve as an allosteric positive modulator of the insect GABA receptor (Tong, 2010).

Certain terpenoid compounds serve as growth inhibitors (Saxena and Tikku, 1990). Certain terpenoids, such as kaurinoic and trachylobanoic acids, have been shown to impede larval growth (Elliger et al., 1976), while others, such as carvacrol, eugenol, farnesol, and geraniol, have been shown to impair insect embryonic development (Mehta et al., 1979).

Side Effects on Non-Target Insects

Many years have passed without conclusive evidence about the effects of plant metabolites on beneficial insects, particularly honeybees (*Apis mellifera*), which are the most important pollinators of cultivated plants (Kumar and Singh 2015). When applied in large quantities, compounds such as andiroba oil, garlic extract, and neem oil could be harmful to honeybee larvae, causing substantial toxicity. In a study, larvae were given syrup containing these oils, and when they emerged as young workers, their body mass was found to be reduced. In addition, studies have shown that garlic extract, neem oil, and rotenone all have the potential to reduce the rate of mobility in adult workers (Ndakidemi et al., 2016). Hence, further research focusing on the impact of secondary plant metabolites on honeybees and other beneficial insects is required.

Prospects of Market Availability of Plant-Based Products and Future Perspectives

Modern intensive farming necessitates effective and eco-friendly insect pest management. Plant secondary metabolites having insecticidal properties are fundamental in environmentally acceptable pest management measures. Wild plant populations include a spectrum of these metabolites, which serve as an effective defense against most harmful insects. Since all valuable plants and their metabolites have not yet been studied thoroughly, further investigations are needed on numerous wild plants to find out their repellant and deterrent properties. In addition, pesticide control can be done alongside the practice of farming. Hence, the incorporation of these effective botanicals must be taken into consideration in future research programmes. With more farmers looking at organic produce, which is costing them more, cost effective methods must be found to manage pests and diseases. In under-developed countries, these products are less affordable without any government support. New socio-economic studies should concentrate on low-cost biopesticides for developing country farmers. Further studies should focus on plant-based products' negative effects on beneficial insects, especially pollinating bees, and natural enemies of agroecosystems.

Nowadays, organic pesticides which are biodegradable, safe end-of-life products are in increasing demand (Badenes-Pérez et al., 2014) and these plant-based pest control products necessitate the cooperation of all stakeholders. Biopesticides are underutilized in agriculture even though numerous international organizations have attempted to increase their use. The government and scientific truths are at odds with the market. Farmers are often bewildered by these goods' poor field persistence. Greater regulation, marketing, and access to safer substances are further benefits. To meet end-user expectations, regulation and industry should aid small and medium-sized firms etc.

Until recently, biopesticides represented only 5% of the crop protection industry. The use of biopesticides grows by roughly 10% every year. Biopesticides are an environmentally benign alternative to synthetic pesticides, and the market for them must continue to expand and these are anticipated to be supplemented by synthetics by the year 2040 (NKWOOD Research 2021).

Plant-based insecticides have been investigated for safety and efficacy for several decades. Current registrations are being altered and amended globally, nationally, and locally. The European Union makes use of botanicals in food supplements. Biopesticides must be registered in comparison to other developed countries in the world. Under the risk-based approach, the active ingredients are assessed for quality, safety, and the environment, as well as regulation 396/2005 on maximum residue limits (MRLs) in food.

Conclusion

Secondary metabolites are produced in a wide variety of ways by plants. Terpenoids are one such family of compounds that have been isolated from plants and have been used for centuries as insecticides to protect agricultural crops. The majority of terpenoids are significant components of essential oils. They have a wide variety of insecticidal properties and, essentially due to their volatility, they are often employed as fumigants. Insecticidal activity is mediated through antifeedants, repellants, and other potential mechanisms that interfere with an insect's fundamental metabolic, biochemical, physiological, and behavioural functions.

Many of the substances have varying degrees of toxicity and the variety in toxicity may be a result of chemical characteristics and structural diversity, both of which contribute significantly to triggering the various harmful effects. For many years, the interactions between terpenoids and pests have been investigated. A better understanding of these complicated connections would pave the way for the future effective use of natural products in rational methods of in pest management.

References

Abbaszadeh, G, Srivastava C., and Walia, S. 2014. Insecticidal and antifeedant activities of clerodane diterpenoids isolated from the Indian bhant tree *Clerodendron infortunatum* against the cotton bollworm, *Helicoverpa armigera*. *J. Insect Sci* 14: 29.

Ali, J.G., Alborn, H.T., and Stelinski, L.L. 2011. Constitutive and induced subterranean plant volatiles attract both entomopathogenic and plant-parasitic nematodes. *J. Ecol* 99, 26-35.

Aslam, I., Kilic, T., Goren, A.C., and Topeu, G. 2006. Toxicity of acetone extract of *Sideritis trojana* and 7-epicandicandiol, 7-epicandicandiol diacetate and 18-acetyl sideroxol against stored pests *Acanthoscelides obtectus* (say), *Sitophilus granarius* (L.) and *Ephestia kuehniella* (Zell). *Indust. Crops and Prod* 23: 171-176.

Badenes-Pérez, F.R., Gershenzon, J., Heckel, D.G. 2014. Insect attraction versus plant defense: young leaves high in glucosinolates stimulate oviposition by a specialist herbivore despite poor larval survival due to high saponin content. *PLoS ONE* 9, e95766.

Balkema-Boomstra, A.G., Zijlstra, S., Verstappen, F.W.A., Inggamer, H., Mercke, P.E., Jongsma, M.A., and Bouwmeester, H.J. 2003. Role of Cucurbitacin C in resistance to spider mite (*Tetranychus urticae*) in cucumber (*Cucumis sativus* L.). *Journal of chemical ecology* 29: 225-235.

Banthorpe, D.V. 1991. Classification of terpenoids and general procedures for their characterization. In B. V. Charlwood and D.V. Banthorpe (Eds.). *Methods in plant biochemistry*. Vol 7 Terpenoids. Academic Press Limited, London, UK. 1-41.

Bian, G., Han, Y., Hou, A., Yuan, Y., Liu, X., Deng, Z., and Liu, T. 2017. Releasing the potential power of terpene synthases by a robust precursor supply platform. *MeTable Eng* 42, 1-8.

Boncan, D.A.T., Tsang, S.S.K., Li, C., Lee, I.T.H., Lam, H.M., Chan, T.F., and Hui, J.H.L. 2020. Terpenes and terpenoids in plants: Interactions with environment and insects. *International Journal of Molecular Sciences* 21: 7382.

Bond,W.J., Lee,W.G., and Craine, J.M. 2004. Plant structural defences against browsing birds: A legacy of New Zealand's extinct moas. *Oikos* 104, 500-508.

Bondi, M.L., Bruno, M., Piozzi, F., Baser, K.H.C., and Simmonds, M.S.J. 2002. Diversity and antifeedant activity of diterpenes from Turkish species of Sideritis. *Biochem. Syst. Ecol* 2002; 28: 299-302.

Born, F.S., Biano, E.M., and da Camana, C.A.G. 2012. Acaricidal and repellent activity of terpenoids from seaweeds collected in Pernambuco, Brazil. *Nat Prod. Commun* 7(4): 463-6.

Brahmkshatriya, P.P., and Brahmkshatriya P.S. 2013. Terpenes: Chemistry, Biological Role, and Therapeutic Applications. In: Ramawat K., Mérillon JM. (eds) *Natural Products.* Springer, Berlin, Heidelberg. https://doi.org/10.1007/978-3-642-22144-6_120.

Bukovinszky, T., Gols, R., Posthumus, M.A., Vet, L.E.M. and van Lenteren, J.C. 2005. Variation in plant volatiles and attraction of the parasitoid *Diadegma semiclausum* (Hellen). *Journal of Chemical Ecology* 31, 461-480.

Byers, K.J., Bradshaw, H., and Riffell, J.A. 2014. Three floral volatiles contribute to differential pollinator attraction in monkeyflowers (*Mimulus*). *J. Exp. Biol* 217, 614-623.

Cascone, P., Iodice, L., Maffei, M.E., Bossi, S., Arimura, G.I., and Guerrieri, E. 2015. Tobacco overexpressing β-ocimene induces direct and indirect responses against aphids in receiver tomato plants. *J. Plant Physiol.* 173, 28-32.

Chaieb, I. 2010. Saponins as insecticides: A review. Tunis. *J. Plant Prot* 5, 39-50.

Coats, J.R., Karr, L.L., and Drewes, C.D. 1991. Toxic and neurotoxic effects of monoterpenoids in insects and earthworms. In: Hedin P (Ed). *Natural occuring pest bioregulators*, ACS Symposium series,Washington D C.

Collins, D.O., Gallimore, W.A., Reynolds, W.F., Williams, L.A.D., and Reese, P.B. 2002. New skeletal sesquiterpenoids, caprariolides A-D, from *Capraria biflora* and their insecticidal activities. *J. Nat. Prod.* 63(11); 1515-1518.

D'Alessandro, M., and Turlings, T.C.J. 2006. Advances and challenges in the identification of volatiles that mediate interactions among plants and arthropods. *The Analyst* 131, 24–32.

De Geyter, E., Lambert, E., Geelen, D., and Smagghe, G. 2007. Novel Advances with plant saponins as natural insecticides. *Pest Technol* 1: 96-105.

Deguenon, J.M., Zhu, J., Denning, S., Reiskind, M.H., Watson, D.W., and Roe, R.M. 2019. Control of filth flies, *Cochliomyia macellaria* (Diptera: Calliphoridae), *Musca domestica* (Diptera: Muscidae) and *Sarcophaga bullata* (Diptera: Sarcophagidae) using novel plant derived methyl ketones. *J. Med. Entomol* 56(6): 1704-1714.

Dicke, M., Gols, R., Ludeking, D., and Posthumus, M.A. 1999. Jasmonic acid and herbivory differentially induce carnivore attracting plant volatiles in lima bean plants. *Journal of Chemical Ecology* 25, 1907-1922.

Dicke, M., van Loon, J.J.A. 2000. Multitrophic effects of herbivore-induced plant volatiles in an evolutionary context. *Entomologia Experimentalis et Applicata* 97:237-249.

Dicke, M., Van Poecke, R.M.P. 2002. Signalling in plant–insect interactions: signal transduction in direct and indirect plant defence. In *Plant Signal Transduction* (eds D. Scheel & C. Wasternack), pp. 289-316. Oxford University Press, Oxford, UK.

Ding, Y., Huffaker, A., Köllner, T.G.,Weckwerth, P., Robert, C.A., Spencer, J.L., Lipka, A.E., and Schmelz, E.A. 2017. Selinene volatiles are essential precursors for maize defense promoting fungal pathogen resistance. *Plant Physiol* 175:1455-1468.

Dubey, S., Manduri, A., Arora, N., and Sankar, P.D. 2014. Transgenics in Phytochemical Defence. *Res. J. Pharm. Biol. Chem. Sci.* 5:763-774.

Dudareva, N., and Pichersky, E. 2006. Metabolic engineering of floral scent of ornamentals. *Journal of Crop Improvement* 18: 325-346.

Duke, S.O. 1991. Plant terpenoids as pesticides. In R F Keeler and A T Tu (Eds), *Toxicology of plant and fungal compounds handbook of natural toxin*, Vol 6. Marcel Dekker Inc, New York, NY.

Elliger, C.A., Zinkel, D.F., Chan, B.G., and Waiss, A.C.J. 1976. Diterpene acids as larval growth inhibitors. *Experientia* 32: 1364-1365.

Fürstenberg-Hägg, J., Zagrobelny, M., and Bak, S. 2013. Plant Defense against Insect Herbivores. *Int. J. Mol. Sci* 14, 10242-10297.

Holopainen, J.K. 2004. Multiple functions of inducible plant volatiles. *Trends in Plant Science* 9: 529-533.

Jawale, C.S. 2014. Larvicidal activity of some Saponin containing plants against the dengue vector *Aedes aegypti. Trend Biotechnol. Res*. 3, 1.

Kazemi, M. 2014. Composition of Essential Oil of *Artemisia indica. J.Environ. Sci* 8: 95-97.

Keeling, C.I., and Bohlmann, J. 2006. Genes, enzymes and chemicals of terpenoid diversity in the constitutive and induced defence of conifers against insects and pathogens. *New Phytol* 170: 657-675.

Khan, A., Islam, M.S., Rahman, M., Zaman, T., and Haque, M.E. 2014. Pesticidal and pest repellency activities of a plant derived triterpenoid 2α, 3β, 21β, 23, 28-penta hydroxyl 12-oleanene against *Tribolium castaneum. Biological research* 47: 68.

Kostyukovsky, M., Rafaeli, A., Gileadi, C., Demchenko, N., and Shaaya, E. 2002. Activation of octopaminergic receptors by essential oil constituents isolated from aromatic plants: Possible mode of action against insect pest. *Pest manag. sci.* 58: 1101-1106.

Kumar, S., and Singh, A. 2015. Biopesticides: present status and the future prospects. *J. Fertil. Pestic* 6: 1-2.

Lee, S., Tsao, R., Peterson, C., and Coats, J.R. 1997. Insecticidal activity of monoterpenoids to western corn rootworm (Coleoptera: Chrysomelidae), two spotted spider mite (Acari: Tetranychidae) and house fly (Diptera: Muscidae). *J. Econ. Entomol* 90(4): 883-92.

Leitner, M., Boland, W., and Mithöfer A. 2005. Direct and indirect defences induced by piercing-sucking and chewing herbivores in *Medicago truncatula. New Phytologist* 167; 597-606.

Li, C.H., Zhang, J.Y., Tuong, T.M.L., Liu, Y., Hoang, X.N., and Gao, J.M. 2020. Cassane diterpenoids from the aerial parts of *Caesalpinia pulcherrima* and their antifeedant and insecticidal activities against *Mythimna separate* and *Plutella xylostella. J. Agric. Food Chem* 68(14): 4227-4236.

Llandres, A.L., Almohamad, R., Brévault, T., Renou, A., Téréta, I., Jean, J., and Goebel, F.R. 2018. Plant training for induced defense against insect pests: A promising tool for integrated pest management in cotton. *Pest Manag. Sci* 74: 2004-2012.

Manchali, S., Murthy, K.N.C., and Patil, B.S. 2011. Crucial facts about health benefits of popular cruciferous vegetables. *J. Fun. Foods* 4(1):94-106.

Mbata, G.N., and Payton, M.E. 2013. Effect of monoterpenoids on oviposition and mortality of *Callosobruchus maculatus* (F.) (Coleoptera: Bruchidae) under hermetic conditions. *J. Stored Prod Res.* 53:43-47.

Mehta, RC. 1979. Inhibition of embryonic development in *Earias vitella* by terpenoids. *Naturwissenschaften* 66: 56.

Mendesil, E. 2014. Plant-resistance to insect herbivores and semiochemicals: Implications for field pea pest management. *Introd. Pap. Fac. Landsc. Archit. Hortic. Crop Prod. Sci* 1:1-37.

Mithöfer, A., and Boland, W. 2012. Plant Defense Against Herbivores: Chemical Aspects. *Annu. Rev. Plant Biol* 63:431-450.

Mordue, A.J., Morgan, E.D., and Nisbet, A.J. 2005. Azadirachtin a natural product in insect control. *Comprehensive molecular insect science* 6: 117-135.

Muller, G.C., Junnila, A., Butler, J., Kravchenko, V.D., Revay, E.E., Weiss, R.W., and Schlein, Y. 2009. Efficacy of the botanical repellents geraniol, linalool and citronella against mosquitoes. *Journal of vector ecology* 34(1): 2-8.

Mumm, R., Schrank, K., Wegener, R., Schulz, S., and Hilker, M. 2003. Chemical analysis of volatiles emitted by *Pinus sylvestris* after induction by insect oviposition. *Journal of Chemical Ecology* 29: 1235-1252.

Ndakidemi, B., Mtei, K., and Ndakidemi, P.A. 2016. Impacts of synthetic and botanical pesticides on beneficial insects. *Agric. Sci.* 7:364-372.

Nesterkina, M., Bernier, U.R., Tubanca, N., and Kravchenko, I. 2018. Repellent activity of monoterpenoid esters with neurotransmitter amino acids against yellow fever mosquito, *Aedes aegypti*. *Open Chem* 16: 95-98.

NKWOOD Research. *Global biopesticides market forecast 2020-2028*. Available online: https://www.inkwoodresearch.com/reports/biopesticides-market/ (accessed on 10 January 2021).

Omar, S., Marcotte, M., Fields, P., Sanchez, P.E., Poveda, L., Mata, R., Jimenez, A., Durst, T., Zhang, J., Mackinnon, S., Leaman, D., Arnason, J.T., and Philogene, B.J.R. 2007. Antifeedant activities of terpenoid isolated from tropical Rutales. *Journal of stored products Research* 43: 92-96.

Pandian, B.A., Sathishraj, R., Djanaguiraman, M., Prasad, P.V., and Jugulam, M. 2020. Role of Cytochrome P450 Enzymes in Plant Stress Response. *Antioxidants* 9:454.

Parry, M.A.J., and Hawkesford, M.J. 2010. Food security: Increasing yield and improving resource use efficiency. *Proc. Nutr. Soc* 69:592-600.

Pezet, J., and Elkinton, J.S. 2014. Hemlock Woolly Adelgid (Hemiptera: Adelgidae) Induces Twig Volatiles of Eastern Hemlock in a Forest Setting. *Environ. Èntomol* 43: 1275-1285.

Pham, T., Chen, H., Yu, J., Dai, L., Zhang, R., and Vu, T.Q.T. 2014. The Differential effects of the blue-stain fungus *Leptographium qinlingensis* on monoterpenes and sesquiterpenes in the stem of Chinese white pine (*Pinus armandi*) saplings. *Forests* 5: 2730-2749.

Plata-Rueda, A., Campos, J.M., Rolim, G.D.S., Martinez, L.C., Dos Santos, M.H., Fernandes, F.L., Serrao, J.E., and Zanuncio, J.C. 2018. Terpenoid constituents of cinnamon and clove essential oils cause toxic effects and behaviour repellency response on granary weevil, *Sitophilus granarius*. *Ecotoxicol. Environ. Saf* 156: 263-270.

Purrington, C.B. 2003. Secondary products: Antifeedant substances in plants. *Encyclopedia of Applied Sciences* 1143-1145.

Saxena, B.P., and Tikku, K. 1990. Impact of natural products on the physiology of phytophagous insects. *Proc. Indian Acad. Sci (Anim Sci)*. 99(3): 185-198.

Senthil-Nathan, S. 2014. Physiological and biochemical effect of neem and other Meliaceae plants secondary metabolites against Lepidopteran insects. *Front. Physiol* 4:359.

Siddiqui, B.S., Afshan, F., Faizi, S., Naqvi, S., Nul, H., and Tariq, R.M. 2002. Two new triterpenoids from *Azadirachta indica* and their insecticidal activity. *J. Nat Pro* 65(8): 1216-1218.

Singh, B., and Sharma, R.A. 2015. Plant terpenes: Defense responses, phylogenetic analysis, regulation and clinical applications. *3 Biotech* 5:129-151.

Solé, J., Sans, A., Riba, M., and Guerrero, A. 2010. Behavioural and electrophysiological responses of the European corn borer *Ostrinia nubilalis* to host-plant volatiles and related chemicals. *Physiol. Åntomol* 35:354-363.

Takabayashi, J., Dicke, M., and Posthumus, M.A. 1994. Volatile herbivore-induced terpenoids in plant-mite interactions: Variation caused by biotic and abiotic factors. *J. Chem. Ecol* 20:1329-1354.

TlakGajger, I., and Dar, S.A. 2021. Plant Allelochemicals as Sources of Insecticides. *Insects* 12, 189. https://doi.org/10.3390/insects12030189.

Tong, F. 2010. *Investigation of mechanism of action of monoterpenoid insecticides on insect gamma-aminobutyric acid receptors and nicotinic acetylcholine receptors*. PhD thesis, IOWA State University, Ames, USA.

Turlings, T.C.J., Bernasconi, M., Bertossa, R., Bigler, F., Caloz, G., and Dorn, S. 1998. The induction of volatile emissions in maize by three herbivore species with different feeding habits: possible consequences for their natural enemies. *Biological Control* 11:122-129.

Vaughan, M.M., Wang, Q., Webster, F.X., Kiemle, D., Hong, Y.J., Tantillo, D.J., Coates, R.M., Wray, A.T., Askew, W., and O'Donnell, C. 2013. Formation of the unusual semi-volatile diterpene rhizathalene by the Arabidopsis class I terpene synthase TPS08 in the root stele is involved in defense against below ground herbivory. *Plant Cell* 25:1108-1125.

Weissteiner, S., Huetteroth, W., Kollmann, M., Weißbecker, B., Romani, R., Schachtner, J., and Schütz, S. 2012. Cockchafer Larvae Smell Host Root Scents in Soil. *PLoS ONE* 7, e45827.

Wink, M. 2009. Mode of action and toxicology of plant toxins and poisonous plants. *Jul. Kühn Archiv*. 421: 93-112.

Wu, H.B., Wang, W.S., Liu, T.T., Qi, M.G., Feng, J.C., Li, X.Y., and Liu, Y. 2016. Insecticidal activity of sesquiterpene lactones and monoterpenoid from the fruits of Carpesium abrontanoides. *Industrial crops and products* 92: 77-83.

Wu, Z., Wei, W., Cheng, K., Zheng, L., Ma, C., and Wang, Y. 2020. Insecticidal activity of titerpenoids and volatile oil from the stems of *Tetraena mongolica*. *Pestic. Biochem. Physiol* 166: 104551.

Wu,W., Liu, F., and Davis, R.W. 2018. Engineering Escherichia coli for the production of terpene mixture enriched in caryophyllene and caryophyllene alcohol as potential aviation fuel compounds. *MeTable Eng. Commun* 6:13-21.

Yang, W., Chen, Xu L, Li, Y., Guo, S., Wang, Z., and Yu, X. 2020. Advances in pharmacological activities of terpenoids. *Natural product communications* 15(3): 1-13.

Biographical Sketch

Bhabesh Deka

Affiliation: Department of Entomology, North Bengal Regional Research and Development Centre, Nagrakata, West Bengal, India

Education: MSc, PhD, PGDBI

Research and Professional Experience: 8 years

Professional Appointments

2021-till date: Scientist-C, Tea Research Association, North Bengal Regional Research & Development Centre, Nagrakata, Jalpaiguri, West Bengal

2014-2021: Scientist-B, Tea Research Association, North Bengal Regional Research & Development Centre, Nagrakata, Jalpaiguri, West Bengal

Honors:

1) Merit award certificate for best graduate from Pub Kamrup College, (Gauhati University) in 2003.

2) Financial assistance for participating in International symposium on Insects, Entomological society of Malaysia (Held on 3-5th Dec, 2012) from Science and Engineering Research Board; DST (Under International Travel support scheme).
3) Research Excellence Award 2020, Institute of Scholars (InSc.)
4) Young Scientist award (8th Academic Achievement Award-2020)
5) Young Scientist Award- 2021 (3rd National Conference on "Recent Advances in Crop Protection including IPM and Environmental Sciences from GLP Perspective" Chennai. Organized by the Dr. B. Vasantharaj David Foundation.)
6) Member of "*Entomological Society of America" Member ID: 80927.*
7) Member of "*European Federation of Biotechnology.*"
8) Member of International Association of Computer Science and Information Technology, Singapore *(IACSIT),* Membership No.: 80344880.
9) Member of Asia-Pacific Chemical, Biological & Environmental Engineering Society *(APCBEES),* Hong Kong, Member No.: 201039.
10) *Editorial Board Member Asian Journal of ConservationBiology* (AJCB) (Scopus). http://www.ajcb.in/archive.php.

Publications from the Last 3 Years:

Paper Published:

1) Deka B, Baruah C, and Babu A. (2021). Entomopathogenic microorganisms: their role in insect pest management. *Egyptian Journal of Biological Pest Control.* 31, 121. https://doi.org/10.1186/s41938-021-00466-7.
2) Deka B and Babu A (2021) Tea Pest Management: A Microbiological Approach. *Applied Microbiology: Open Access.* 7(8):1-9.
3) Deka B, Babu A, Baruah C and Barthakur M. (2021). Nanopesticides: a systematic review of their prospects with special reference to tea pest management. *Frontiers in nutrition (special issue on Tea (Camellia sinensis L): Cultivation to the Cup That Cheers).* doi: 10.3389/fnut.2021.686131.
4) Deka B., Tomberlin JK., and Babu A. (2021): Insects as food and feed: a promising agricultural sector with special reference to India.

Journal of Insects as Food and Feed. 7 (4): 471 – 482. https://doi.org/10.3920/JIFF2020.0131.

5) Deka B, Pandey AK, Babu A, Baruah C & Sarkar S (2021): Acaricidal and ovicidal properties of *Lippia alba*essential oil and its chemical constituents against red spider mite, *Oligonychus coffeae* Nietner (Acari: Tetranychidae) infesting tea crops. *Archives of Phytopathology and Plant Protection.* DOI: 10.1080/03235408.2021.1938354.
6) Sarkar S., Babu A., Chakraborty K., Deka B. & Roy S. (2021): Seasonal abundance of *Cotesia ruficrus* (Hymenoptera: Braconidae) and its host tea looper, *Hyposidra talaca* (Lepidoptera: Geometridae) in tea ecosystem. *International Journal of Pest Management.* DOI: 10.1080/09670874.2021.1918359.
7) Sarkar S., Babu A., Chakraborty K., Deka B., Kashyap B., and Adhikary B. (2021): Study on varietal preference and nutritional indices of tea looper *Hyposidra talaca* (Lepidoptera: Geometridae) on ten selected tea varieties. *International Journal of Tropical Insect Science.* https://doi.org/10.1007/s42690-021-00502-x.
8) Pandey A.K., Deka B., Varshney R., Cheramgoi E.C. and Babu A. (2021) Do the beneficial fungi manage phytosanitary problems in the tea agro-ecosystem? *BioControl.* https://doi.org/10.1007/s10526-021-10084-9.
9) Subramaniam M.S.R., Babu A. and Deka B. (2021): *Lecanicillium lecanii* (Zimmermann) Zare & Gams, as an efficient biocontrol agent of tea thrips, *Scirtothrips bispinosus* Bagnall (Thysanoptera: Thripidae). *Egyptian Journal of Biological Pest Control.* 31:38. https://doi.org/10.1186/s41938-021-00380-y.
10) Deka B., Baruah C. and Barthakur M. (2021). In silico tertiary structure prediction and evolutionary analysis of two DNA-binding proteins (DBP-1 and DBP-2) from *Hyposidra talaca* nucleopolyhedrovirus (HytaNPV). *Biologia.* https://doi.org/10.2478/s11756-020-00665-x.
11) Sarkar S., Babu A., Chakraborty K., Deka B. and Roy S. (2021) *Eocanthecona furcellata* (Wolff) (Hemiptera: Pentatomidae), a potential biocontrol agent of the black inch worm, *Hyposidra talaca* Walker (Lepidoptera: Geometridae) infesting tea. *Phytoparasitica.* 49(3), 363-376. https://doi.org/10.1007/s12600-021-00888-x.
12) Deka B., Babu A. and Sarkar S. (2020). *Scirtothrips dorsalis*, Hood (Thysanoptera: Thripidae): A major pest of tea plantations in North

East India. *Journal of Entomology and Zoology Studies*. 8(4): 1222-1228.

13) Sarkar S., Babu A., Chakraborty K. and Deka B. (2020). Biology and life history of *Cotesia ruficrus* (Hymenoptera: Braconidae) a potential parasitoid of *Hyposidra talaca* (Lepidoptera: Geometridae) larvae, a major tea pest. *Journal of Biopesticides*, 13(1):79-84.
14) Kumhar K.C., Babu A., Arulmarianathan J.P., Deka B., Bordoloi M., Rajbongshi H. and Dey P. (2020). Role of beneficial fungi in managing diseases and insect pests of tea plantation. *Egyptian Journal of Biological Pest Control*. 30:78. https://doi.org/10.1186/s41938-020-00270-9.
15) Sarkar S., Babu A., Chakraborty K. and Deka B. (2020): Predatory behaviour of *Eocanthecona furcellata* (Hemiptera: Pentatomidae) fed on larvae of *Hyposidra talaca* (Lepidoptera: Geometridae), a major tea pest of North East India and rearing on an alternative host. *International Journal of Scientific Research*. 9 (3) ISSN No. 2277-8179, doi: 10.36106/ijsr.
16) Deka B., Babu A., Sarkar S., Kanrar B., Patanjali P.K. and Bhandari P. (2020): Role of *Prosopis juliflora* as a botanical synergist in enhancing the efficacy of neem kernel powder against tea red spider mites, *Oligonychus coffeae* (Acari: Tetranychidae). *SN Applied Sciences* (Springer nature). https://doi. org/ 10.1007/s42452-020-2354-7.
17) Roy S., Rahman A., Babu A. and Deka B. (2019): Integrated Pest Management Strategies for Scale Insects.*Two & a bud*, 65: 1-2.
18) Sarkar S., Babu A., Chakraborty K. and Deka B. (2019): Study on the Biology, Feeding Behaviour and predatory Potential of *Sycanus collaris* (Fabricius) (Heteroptera: Reduviidae), a new Predator of *Hyposidra talaca* (Walk.) (Lepidoptera: Geometridae), a major Tea pest and mass Rearing on *Corcyra cephalonica* (Stainton) in Laboratory', *International Journal of Current Advanced Research*, 08(06), pp. 19258-19262. DOI: http://dx.doi.org/10.24327/ijcar.2019.19262.3705.
19) Nguyen T.T., Suryamohan K., Kuriakose B., Janakiraman V., Reichelt M., Chaudhuri S., Guillory J., Divakaran N., Rabins P.E., Goel R., Deka B., Sarkar S., Ekka P., Tsai Y., Vargas D., Santhosh S., Mohan S., Chin C., Korlach J., Thomas G., Babu A. & Seshagiri S. (2018): Comprehensive analysis of single molecule sequencing-derived complete genome and whole transcriptome of *Hyposidra*

talaca nuclear polyhedrosis virus. *Scientific Reports-Nature 8:8924*| doi: 10.1038/s41598-018-27084-y.

Book Chapters:

1) Deka B. and Babu A. 2021. Thrips (*Scirtothrips dorsalis*, Hood): *Vectors of Tospoviruses in Agricultural Crops. Essentials of Economic Entomology*. AkiNik Publications. ISBN: 978-93-90541-81-2. Volume: 2. 35-53. DOI: https://doi.org/10.22271/ ed.book.1039
2) Bora D.S., Deka B., and Fahmi F. 2012: Response of *Exorista sorbillans* (Diptera: Tachinidae) to Volatiles of *Persea bombycina* Kost, a Host Plant of *Antheraea assamensis* Helfer. *Proceedings of International conference on Bioresources and Human sustenance.*
3) Deka B. and Baruah C. 2021. Prospects of Smart farming in India: a key to sustainable agricultural development. *Artificial Intelligence and Smart Agriculture: Technology and Applications*. CRC Press.
4) Babu A, Deka B and Roy S. 2021. Integrated pest management in plantation crops with special reference to tea. *Improving Integrated Pest Management for Crop Protection and Food Security*. IGI Global.

Archana Borah

Affiliation: Gauhati University

Education: MSc, PhD

Research and Professional Experience:

2020-till date	Assistant Professor, B. Borooah College, Guwahati, Assam.
2019-2020	Assistant Professor, University of Science and Technology, Meghalaya.

Honors:

1) PhD with specialization in biochemistry.
2) Life Member of Zoological Society of Assam.

Publications from the Last 3 Years:

1) Archana Borah: *Overview of materials and techniques for encapsulation of natural products: A mini review* accepted in International Journal of pharmaceutical science and research.

Uma Dutta

Affiliation: Biology Lab, Department of Zoology, Cotton University, Pan Bazar, Ghy-1

Education: MSc, LLB, PhD

Research and Professional Experience: 16 years of research experience and 29 years of teaching experience

Professional Appointments:

2020-till date: Head, Dept. of Zoology, Cotton University
2006-2009: Associate professor, Dept. of Zoology, Cotton University
2001-2006: Senior lecturer: Dept. of Zoology, Cotton College & Halflong Govt. College

Honors:

1) National scholarship (from HSLC to PG level): 1984-1990
2) Best paper presentation award in STOX India Conference: 1996
3) UGC Fellowship Award for Ph.D. degree: UGC Fellowship Award for Ph.D. degree
4) UGC Travel Award to present a paper in International Congress of Toxicology in Barcelona, Spain: 2010
5) EUROTOX 2011 Fellowship Award (Senior Scientist Group): 2011
6) UGC Travel Award to present a paper in EUROTOX 2013 in Interlaken, Switzerland: 2013

Publications from the Last 3 Years:

1) Sosmitha Girisa, Sahu Henamayee, Dey Parama, Varsha Rana, Uma Dutta & Ajaikumar B. Kunnumakkara "Targeting Farnesoid X receptor (FXR) for developing novel therapeutics against cancer"; *Molecular Biomedicine* vol. 2, no. 21 (2021); Nature. https://doi.org/10.1186/s43556-021-00035-2
2) Girisa S, Rana V, Parama D, Dutta U, Kunnumakkara AB. "Differential roles of farnesoid X receptor (FXR) in modulating apoptosis in cancer cells.", *Advances in Protein Chemistry and Structural Biology.* 2021; 126:63-90. DOI:10.1016/bs. apcsb.2021.02.006.
3) Semim Akhtar Ahmed, Dey Parama, Enush Daimari, Sosmitha Girisa, Kishore Banika, Choudhar Harsha, Uma Dutta, AjaikumarB.Kunnumakkaraa "Rationalizing the therapeutic potential of apigenin against cancer", *Life Sciences*; Volume 267, 2021, 118814; https://doi.org/10.1016/j.lfs.2020.118814
4) Ajaikumar B. Kunnumakkara, Varsha Rana, Dey Parama, Kishore Banik, SosmithaGirisa, SahuHenamayee, Krishan Kumar Thakur, Uma Dutta, Prachi Garodia, Subash C. Gupta, Bharat B. Aggarwal, "COVID-19, cytokines, inflammation, and spices: How are they related?" *Life Sciences*, 2021, 119201, ISSN 0024-3205, https://doi.org/10.1016/j.lfs.2021.119201.
5) Dey Parama, Monikongkona Boruah, Kumari Yachna, Varsha Rana, Kishore Banik, Choudhary Harsha, Krishan Kumar Thakur, Uma Dutta, Aditya Arya, Xinliang Mao, Kwang Seok Ahn, Ajaikumar B. Kunnumakkara, "Diosgenin, a steroidal saponin, and its analogs: Effective therapies against different chronic diseases", *Life Sciences* 2020, vol. 260, 118182, https://doi.org/10.1016/j.lfs.2020.118182.
6) Nabanita Medhi, Uma Dutta. "Protective effect of Aloe vera and *Bryophyllumpinnatum* extracts against Smokeless tobacco induced nephrotoxicity in albino mice" *Asian Journal of Pharmacy and Pharmacology* 2019; 5(5):901-908, DOI https://doi.org/10.31024/ajpp.2019.5.5.7
7) Uma Dutta, Shraddha Basu, Manidipa Baruah, "Faith in spirituality to treat infertility: An epidemiological study in certain parts of northern India", vol no. 40, pp 442-465, *Asian Association of Indigenous and Cultural Psychology (AAICP), International*

Conference Proceedings, 25th -27th July 2018, Kota Kinabalu, Sabah, Malaysia.

8) Uma Dutta, "Bio-assay of chemoprotective role of shoot of *Bambusabalcooa*Roxb. and its isolated flavonoid in Ethephon-induced oxidative stress in the sub-cellular pulmonary damage of albino rat *Plant Biodiversity and Applications*, CODE- OP-PBA007, pp 134-135, Proceedings of BIODIVERSE 2018.
9) Bidisha Kakati and Uma Dutta; "Study of seasonal variation in butterfly diversity in certain park areas of Tezpur, Assam", *Insect Biodiversity*, CODE- PP-INB007, pp 107-108, Proceedings of BIODIVERSE 2018.
10) Kasturi Dutta and Uma Dutta, "A comparative study on butterfly diversity in the heart area-Nehru Park and Outskirt area – IIT Campus of Guwahati City, Assam", *Insect Biodiversity*, CODE- PP-INB006, pp 105-106, Proceedings of BIODIVERSE 2018.
11) Jutikana Mishra and Uma Dutta, "Enthno-traditional usage of diversity of plant species for the treatment of menstrual disorders in certain villages of Tezpur subdivision, Assam, India" *Human and Linguistic Diversity*. CODE- PP-HLD015, pp 93-94. Proceedings of BIODIVERSE 2018.
12) Deepshika Moran and Uma Dutta, "Efficacy of methanolic exract of *Alternanthera sessilis* (L.) in combination with the venom of *Polistes flavus* on nicotine derived nitrosomine ketone (NNK) induced pulmonary carcinoma in albino mice" *Animal Biodiversity*, CODE-PP-ANB004, pp 3-4 Proceedings of BIODIVERSE 2018.

Azariah Babu

Affiliation: Tea Research Association, North Bengal Regional Research and Development Centre, Nagrakata, West Bengal, India

Education: MSc, PhD

Research and Professional Experience: 28 years

Professional Appointments:

2012-till date	Chief Scientist & Deputy Director (R), Tea Research Association, North Bengal Regional Research and Development Centre, Nagrakata, West Bengal, India
2011-2012	Head, Department of Entomology, Tea Research Association. Tocklai Experimental Station, Jorhat, Assam
2002-2011	Head, Division of Entomology, UPASI Tea Research Foundation & Tea Research Institute, Valparai.
1998-2002	Scientist, Entomology Research Institute, Loyola College, Chennai-34.

Honors:

1) Visited Sri Lanka and presented a paper on "Management of social forestry biodiversity: An entomological approach" at the Regional Session on Global Biodiversity Forum (South and South East Asia) at Sri Lanka.
2) Awarded Young Scientist (under Scheme for Young Scientists (SYS), on Bionomics and tritrophic interactions of phytophagous Coccinellids by the Department of Science and Technology (DST), Government of India, New Delhi (1999-2001).
3) Visiting Scientist at the Swiss Federal Institute of Technology, Zurich, [Applied Entomology] and worked on the Research Topic *Headspace volatiles of infested and un-infested cowpeas: characterization and bioassay*.
4) Recognized examiner for evaluation of PhD thesis of Calicut University and Annamalai University, Bharathiar University and Madras University.
5) Recognized Research Guide for PhD programme in Entomology – Zoology Bharathiar University Coimbatore the Dibrugargh University and Gauhati University, Assam
6) Lifetime Achievement Award-2021 in appreciation of "Outstanding contribution toTea entomology and Higher Education" by the Dr. B. Vasantharaj David Foundation Chennai.

Publications from the Last 3 Years:

Paper Published:

1) Deka B, Baruah C, and Babu A. (2021). Entomopathogenic microorganisms: their role in insect pest management. *Egyptian Journal of Biological Pest Control.* 31, 121. https://doi.org/10.1186/s41938-021-00466-7.
2) Deka B and Babu A (2021) Tea Pest Management: A Microbiological Approach. *Applied Microbiology: Open Access.* 7(8):1-9.
3) Deka B., Tomberlin JK., and Babu A. (2021): Insects as food and feed: a promising agricultural sector with special reference to India. *Journal of Insects as Food and Feed.* 7 (4): 471 – 482. https://doi.org/10.3920/JIFF2020.0131.
4) Deka B, Pandey AK, Babu A, Baruah C & Sarkar S (2021): Acaricidal and ovicidal properties of *Lippia alba*essential oil and its chemical constituents against red spider mite, *Oligonychus coffeae* Nietner (Acari: Tetranychidae) infesting tea crops. *Archives of Phytopathology and Plant Protection.* DOI: 10.1080/03235408.2021.1938354.
5) Sarkar S., Babu A., Chakraborty K., Deka B. & Roy S. (2021): Seasonal abundance of *Cotesia ruficrus* (Hymenoptera: Braconidae) and its host tea looper, *Hyposidra talaca* (Lepidoptera: Geometridae) in tea ecosystem. *International Journal of Pest Management.* DOI: 10.1080/09670874.2021.1918359.
6) Sarkar S., Babu A., Chakraborty K., Deka B., Kashyap B., and Adhikary B. (2021): Study on varietal preference and nutritional indices of tea looper *Hyposidra talaca* (Lepidoptera: Geometridae) on ten selected tea varieties. *International Journal of Tropical Insect Science.* https://doi.org/10.1007/s42690-021-00502-x.
7) Pandey A.K., Deka B., Varshney R., Cheramgoi E.C. and Babu A. (2021) Do the beneficial fungi manage phytosanitary problems in the tea agro-ecosystem? *BioControl.* https://doi.org/10.1007/s10526-021-10084-9.
8) Subramaniam M.S.R., Babu A. and Deka B. (2021): *Lecanicillium lecanii* (Zimmermann) Zare & Gams, as an efficient biocontrol agent of tea thrips, *Scirtothrips bispinosus* Bagnall (Thysanoptera: Thripidae). *Egyptian Journal of Biological Pest Control.* 31:38. https://doi.org/10.1186/s41938-021-00380-y.

9) Sarkar S., Babu A., Chakraborty K., Deka B. and Roy S. (2021) *Eocanthecona furcellata* (Wolff) (Hemiptera: Pentatomidae), a potential biocontrol agent of the black inch worm, *Hyposidra talaca* Walker (Lepidoptera: Geometridae) infesting tea. *Phytoparasitica.* 49(3), 363-376. https://doi.org/10.1007/s12600-021-00888-x.
10) Deka B., Babu A. and Sarkar S. (2020). *Scirtothrips dorsalis*, Hood (Thysanoptera: Thripidae): A major pest of tea plantations in North East India. *Journal of Entomology and Zoology Studies.* 8(4): 1222-1228.
11) Sarkar S., Babu A., Chakraborty K. and Deka B. (2020). Biology and life history of *Cotesia ruficrus* (Hymenoptera: Braconidae) a potential parasitoid of *Hyposidra talaca* (Lepidoptera: Geometridae) larvae, a major tea pest. *Journal of Biopesticides*, 13(1):79-84.
12) Kumhar K.C., Babu A., Arulmarianathan J.P., Deka B., Bordoloi M., Rajbongshi H. and Dey P. (2020). Role of beneficial fungi in managing diseases and insect pests of tea plantation. *Egyptian Journal of Biological Pest Control.* 30:78. https://doi.org/10.1186/s41938-020-00270-9.
13) Sarkar S., Babu A., Chakraborty K. and Deka B. (2020): Predatory behaviour of *Eocanthecona furcellata* (Hemiptera: Pentatomidae) fed on larvae of *Hyposidra talaca* (Lepidoptera: Geometridae), a major tea pest of North East India and rearing on an alternative host. *International Journal of Scientific Research.* 9 (3) ISSN No. 2277-8179, doi: 10.36106/ijsr.
14) Deka B., Babu A., Sarkar S., Kanrar B., Patanjali P.K. and Bhandari P. (2020): Role of *Prosopis juliflora* as a botanical synergist in enhancing the efficacy of neem kernel powder against tea red spider mites, *Oligonychus coffeae* (Acari: Tetranychidae). *SN Applied Sciences (Springer nature).* https://doi. org/ 10.1007/s42452-020-2354-7.
15) Roy S., Rahman A., Babu A. and Deka B. (2019): Integrated Pest Management Strategies for Scale Insects.*Two & a bud*, 65: 1-2.
16) Sarkar S., Babu A., Chakraborty K. and Deka B. (2019): Study on the Biology, Feeding Behaviour and predatory Potential of *Sycanus collaris* (Fabricius) (Heteroptera: Reduviidae), a new Predator of *Hyposidra talaca* (Walk.) (Lepidoptera: Geometridae), a major Tea pest and mass Rearing on *Corcyra cephalonica* (Stainton) in Laboratory', *International Journal of Current Advanced Research*,

08(06), pp. 19258-19262. DOI: http://dx.doi.org/10.24327/ijcar.2019.19262.3705.

17) Nguyen T.T., Suryamohan K., Kuriakose B., Janakiraman V., Reichelt M., Chaudhuri S., Guillory J., Divakaran N., Rabins P.E., Goel R., Deka B., Sarkar S., Ekka P., Tsai Y., Vargas D., Santhosh S., Mohan S., Chin C., Korlach J., Thomas G., Babu A. & Seshagiri S. (2018): Comprehensive analysis of single molecule sequencing-derived complete genome and whole transcriptome of *Hyposidra talaca* nuclear polyhedrosis virus. *Scientific Reports-Nature 8:8924*| doi: 10.1038/s41598-018-27084-y.

Book Chapters:

1) Deka B. and Babu A. 2021. Thrips (*Scirtothrips dorsalis*, Hood): Vectors of Tospoviruses in Agricultural Crops. *Essentials of Economic Entomology.* AkiNik Publications. ISBN: 978-93-90541-81-2. Volume: 2. 35-53. DOI: https://doi.org/10.22271/ed.book.1039\

2) Babu A, Deka B and Roy S. 2021. Integrated pest management in plantation crops with special reference to tea. *Improving Integrated Pest Management for Crop Protection and Food Security.* IGI Global.

Chapter 5

Sulfuric Acid Promoted Molecular Rearrangement: Applications in Synthesis of Terpenoid Compounds

Ajoy K. Banerjee[1,*], Lisbeth Mendoza[1], Betzabeth Briceño[1], William J. Vera[1], Alexis Maldonado[1], Elvia V. Cabrera[2] and Dioni Arriechi[3]

[1]Chemistry Center, Venezuelan Institute of Scientific Research (IVIC), Caracas, Venezuela
[2]Faculty of Chemical Engineering Central University of Ecuador, Quito, Ecuador
[3]Department of Chemistry, Technical University Federico Santa Maria, Valparaiso, Chile

Dedicated with admiration and affection to the memory of Prof. Alex Nickon, Johns Hopkins University, Baltimore, USA

Abstract

The organic compounds containing functional groups like ketone (saturated, unsaturated), alcohol (primary, secondary, tertiary), epoxide etc. undergo rearrangement with sulfuric acid, yielding unexpected products. Alkyl groups attached to the aromatic ring migrate with sulfuric

* Corresponding Author's E-mail: aabanerje@gmail.com.

In: Terpenes and Terpenoids
Editor: Charles A. Davies
ISBN: 978-1-68507-559-0

acid from one position to another. Several N-phenyl-hydroxyamines rearrange with sulfuric acid to amino phenols. The oximes with sulfuric acid furnish substituted amides and lactams.

Keywords: sulfuric acid, alcohol, epoxide, ketone, diol

Introduction

For a long time during the syntheses of organic compounds and natural products, many unexpected transformations of organic compounds with sulfuric acid have been observed. The information obtained which is scattered in several books and journals have been brought together for the first time in the form of a short review article. The organic chemists had to develop mechanisms to explain the formation of these unexpected products which have provided additional information to our concept and understanding of the mechanism of organic chemistry. Some of these rearrangement products have been utilized for the synthesis of terpenoid compounds. The formation of phenols are observed by the treatment of dienones with sulfuric acid. Many epoxides are cleaved with sulfuric acid and yield ketones. Several monocyclic ketones undergo rearrangement with sulfuric acid and some of these products proved useful for the synthesis of monoterpenes. The ring cleavage with sulfuric acid has been observed in compounds containing saturated and α,β-unsaturated carbonyl groups. Many compounds obtained by acid catalyzed rearrangement have not yet received any application and can be utilized in near future for the synthesis of terpenoid compounds (Heathcock 1973; Heathcock et al., 1983; Pirrung, Morehead Jr. and Young 2000) by using the well-known organic reactions. The acid catalyzed rearrangements of terpenoid compounds have been recorded (King and De Mayo, 1964). The mechanisms of the well-known reactions have not been discussed.

Rearrangement of Ketone

Many ketones including α,β-unsaturated ketones and dienones undergo interesting rearrangement with sulfuric acid. A few examples are given below:

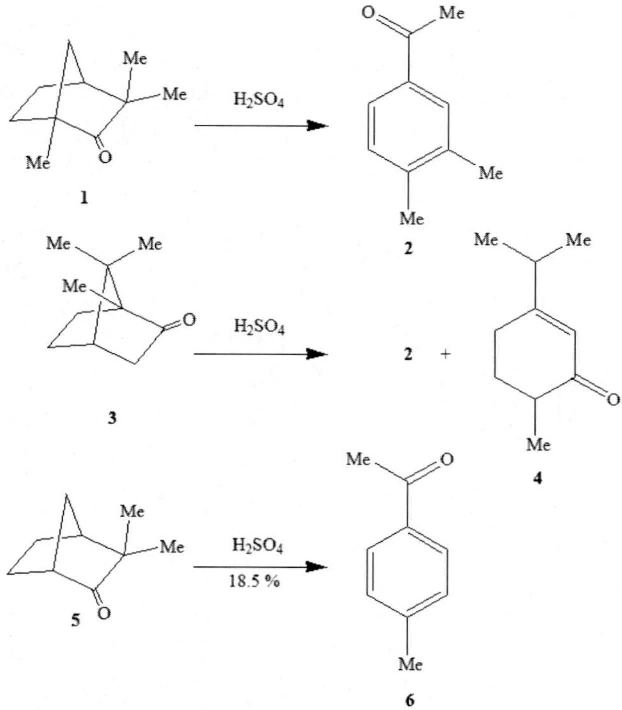

Scheme 1. Rearrangement of ketones 1, 3 and 5 to acetophenone 2, unsaturated ketone 4 and acetophenone 6 respectively.

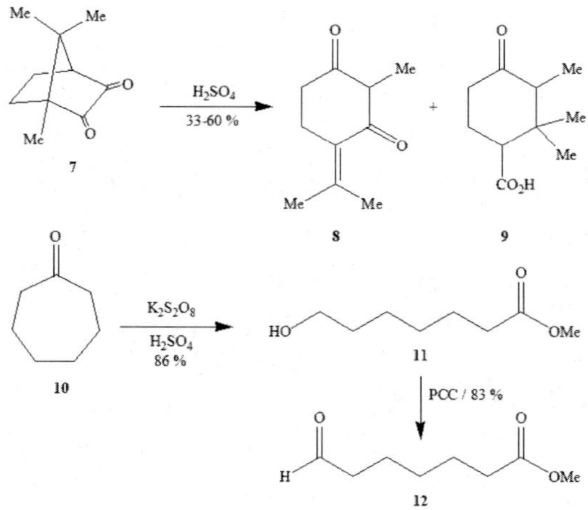

Scheme 2. The rearrangement of camphorquinone **7** and cycloheptanone **10** to isocamphorquinone **8** and 7-hydroxy heptanoate **11** respectively.

Fenchone 1 on treatment with cold conc. sulfuric acid at 80-100°C affords (Marsh 1899) 3.4-dimethylacetophenone **2**. Under similar condition camphor **3** yields (Armstrong and Kippings 1893) acetophenone **2** and carvenone **4**. Camphenilone **5** rearranges (Noyce 1950) to *p*-methylacetophenone **6** (**Scheme 1**). The mechanisms of these rearrangements have been discussed (Noyce 1950, Luz and Roberts 1962 and thus require no detail.

The camphorquinone **7** rearranges to isocamphorquinone **8** and keto-carboxylic acid **9** with cold fuming sulfuric acid (Levine 1960). Cycloheptanone **10** with three equivalents of potassium persulfate in methanol and sulfuric acid (Ballini, Marcantoni and Petrini 1991, Canonica, Ferrari and Sisti, 1989) yields methyl 7-hydroxy heptanoate **11** (**Scheme 2**) in 86% yield which is converted to aldehyde **12** in 83% yield by oxidation with pyridinium chlorochromate, a key intermediate in the synthesis of prostanoids (Elison 1973).

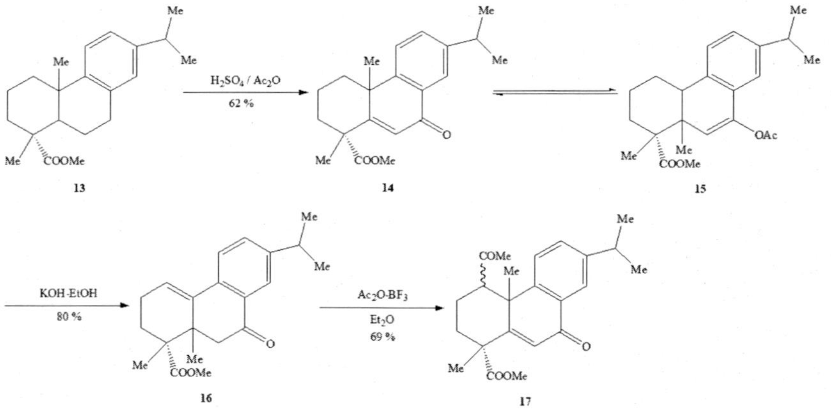

Scheme 3. Synthesis of ketone **17** by the rearrangement of dehyroabietic acid ester **13** with sulfuric acid.

A very interesting rearrangement of α,β-unsaturated ketone **14**, derived from dehydroabietic acid ester **13**, with sulfuric acid and acetic anhydride at room temperature has been recorded (Tahara, Miizuno and Ohsawa 1972). An enolacetate **15** accompanied with C-10 methyl migration of the C-5 position (**Scheme 3**) is formed. Alkaline hydrolysis (KOH-EtOH) of 15 at room temperature gives ketone **16** which on acetylation yield **17**.

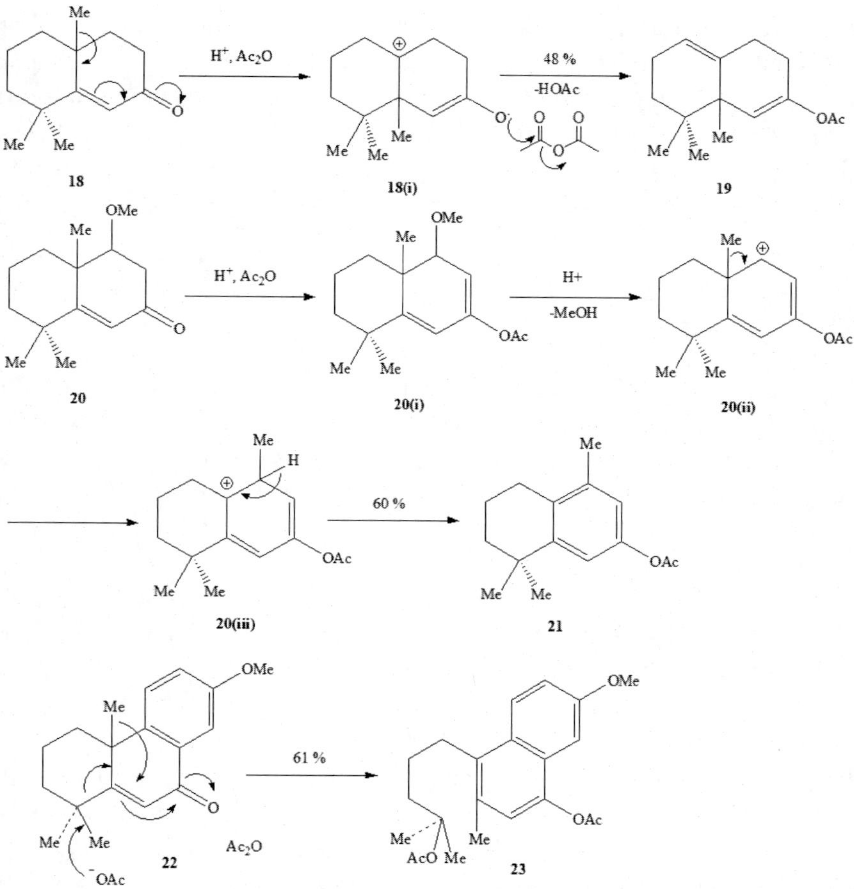

Scheme 4. Rearrangement of α,β-unsaturated ketones **18, 20** and **22** to acetate **19, 21** and **23** respectively.

The ketone **18** with sulfuric acid and acetic anhydride (**Scheme 4**) (Banerjee, Carrasco and Peña 1990) causes angular methyl migration and formation of enolacetate leading the formation of the cation **18(i)** through the path depicted. Finally the cation is converted to the diene **19**. The ketone **20** on similar treatment affords aromatic acetate **21** via the enol acetate **20(i)** which eliminates methanol and forms the carbocation **20(ii)** followed by the formation of cation **20(iii)** that leads the formation of the aromatic acetate **21**.

The ketone **22** under similar reaction condition forms the enol acetate, undergoes angular methyl migration, aromatization and ring scisson respectively under the reaction condition to yield the diacetate **23 (Scheme 4)**. The functional groups present in the compounds **19** and **21** lead us to believe that these compounds can be possibly utilized for the synthesis of bicyclic sesquiterpenes (Heathcock 1973) like thujopsene, drimenal etc.

Rearrangement of Dienones

Several dienones exhibit interesting transformations with sulfuric acid. The dienone **24** on treatment with 50% sulfuric acid undergoes aromatization (Kropp 1963) leading to the formation of phenols 25 (35%), 2**6** (27%) and **27** (20%) (Scheme 5). We believe that the phenols (**25-27**) can be utilized for the synthesis of bicyclic sesquiterpenes (Graham et al., 1983) platphyllide, rishitinol etc. The rearrangement of the dienone **28** with sulfuric acid and acetic anhydride (Arnold, Buckley and Richter 1947; Marvell and Geiszler 1952; Marvell and Stephenson 1955) yields the acetate followed by methyl migration and aromatization to lead to the formation of the acetate **29** (**Scheme 5**). The route to this transformation has been indicated by the arrows. The dienone **30** on similar treatment affords the acetate **31**. The rearrangement whose mechanism has been discussed is interesting (Marvell and Geisler 1952). The acetates **29** and **31** can be converted to quinones (Thompson 1992).

The dienone **32** with sulfuric acid and acetic anhydride followed by alkaline hydrolysis is converted into the phenol 33 (80%) (Woodward and Singh 1950) whereas the dienone 32 with aqueous mineral acid yields the phenol **33** (20%) and phenol 34 (80%) respectively (Scheme 6). The phenol **33** can be utilized by using standard organic reactions to synthesize the bicyclic sesquiterpenes (Pirrung, Morehead Jr and Young 2000) like 4-hydroxy isocadalene, cinabicol and many others. The methyl derivative of **34** has been chosen as the starting material for the synthesis of sesquiterpene alkaloid dendobrine (Hayakawa et al., 1971). The aromatization of the dienone **35** with sulfuric acid and acetic anhydride produces (Mclamore. 1951) triacetate **36** (**Scheme 6**) which has proved useful for the hydroxybenzoquinone.

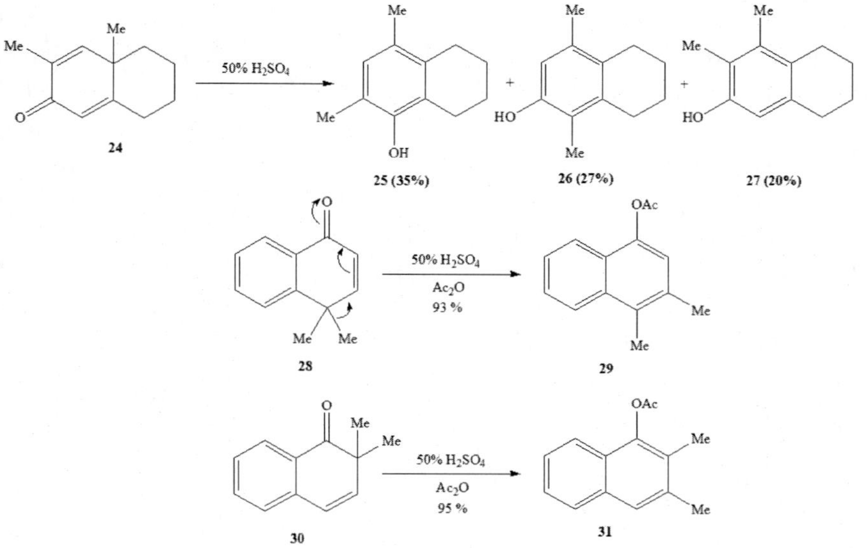

Scheme 5. Aromatization of dienones **24, 28** and **30** to phenol **(25, 26, 27)** and aromatic acetates **29 and 31** respectively.

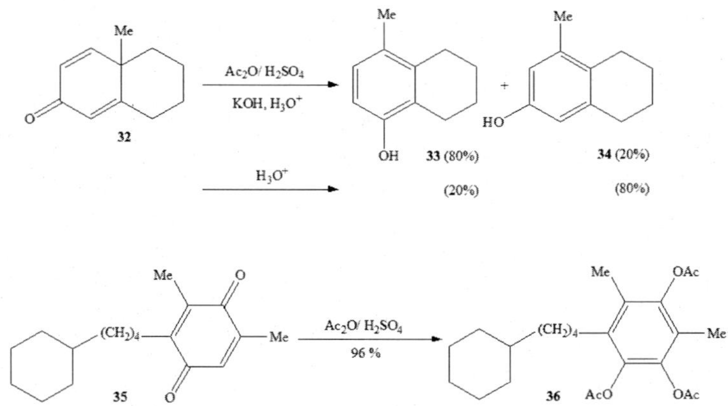

Scheme 6. The aromatization of the dienones **32** and **35** with sulfuric acid to phenol **33, 34** and triacetate **36** respectively.

Rearrangement of Epoxides

Many epoxides undergo cleavage with acids leading to the formation of ketone. Banerjee and collaborators (Cabrera, Sanchez and Banerjee 2011)

have reported that the formation of the tetralone **39** in 53% yield (Scheme 7) by treatment of the epoxide of the teralene **38**. The tetralone **39** is an intermediate for morphine (Soffer et al., 1950) and steroids (Cardwell et al., 1951). The dihydrotetralene **38** is obtained by heating tetralone 37 by heating with 2.4-pentanediol and a catalytic amount of
p-toluenesulfonic acid. The same procedure is applied for the conversion of the tetralone **40** to yield the tetralone **42** via tetralene **41**. The tetralone **42** has been utilized for the synthesis sesquiterpenes occidol (Vera and Banerjee 2006) and emmotin G methylether (Banerjee and Vera 2004).

The tetralone **44**, prepared (Cabrera et al., 2014) from the tetralone **43**, is converted to alkene **45** by the procedure described above (Cabrera, Sanchez and Banerjee, 2011). Epoxidation of the alkene **45** followed by its rearrangement with sulfuric acid (10%) gives the tetralone **46**, which is a potential intermediate for the diterpene carnosic acid. The acid is known to possess antioxidant activity as well as anticancer activity and antiviral properties.

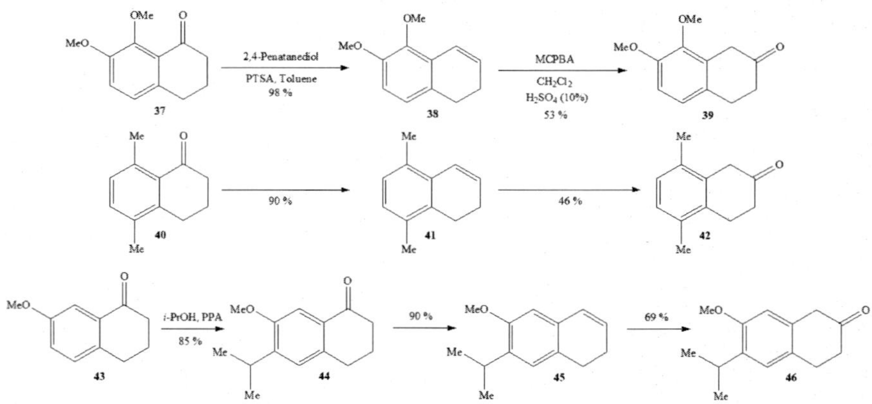

Scheme 7. The transformation of α-tetralones **37, 40** and **43** to tetralones **39, 42** and **46** respectively.

The epoxides **47** (Kato et al., 1990) and **48** (Thompson and Heatchcock 1992) on treatment with sulfuric acid suffer interesting rearrangement and yield the bromotropanone **49** (unspecified yield) and alcohol **50** (70%) respectively. The compound **50** is an intermediate for the synthesis of sesquiterpene lactone.

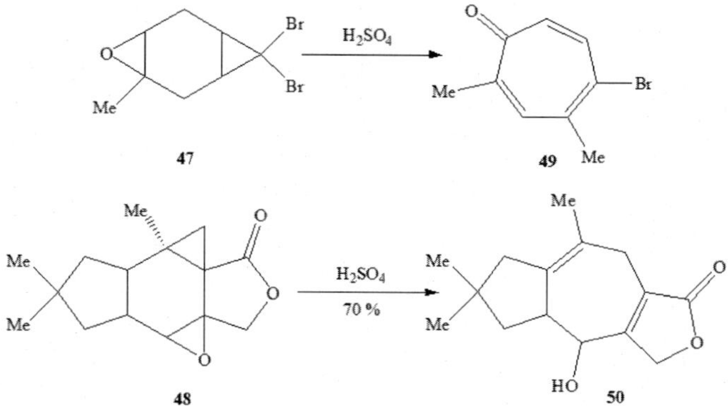

Scheme 8. Rearrangement of epoxides **47** and **48** to bromotropinone **49** and hydroxyl lactone **50** respectively.

Rearrangement of Alcohols

Several alcohols (primary, secondary and tertiary) on treatment with sulfuric acid undergo novel rearrangement and provide interesting products. The alcohol with sulfuric acid leads to the formation of carbocation which can further react in several ways – e.g., direct elimination or substitution reaction depending upon the nucleophilic species present, to give a more stable product.

However, Wagner-Meerwein rearrangement (Streitwiser 1956; Hogeween and vankruchen 1979) will occur if it leads to new carbocation being more stable than the original. The carbocation may undergo a 1.2-alkylshift, generating a different carbocation which suffers further rearrangement by many paths – e.g., by loss of proton to yield an alkene.

The following examples indicate the rearrangement of alcohols with sulfuric acid. Isoborneol **51** (Sorensen 1976), alcohol **52** (Corey and Nozoe 1965) and the alcohol **53** (Kagabu and Kojima 1992) with sulfuric acid yield camphene **54**, α-caryophyllene alcohol **55** and indane **56** respectively (**Scheme 9**). The indane **56** has been converted to the commercial perfume celestolide **57** by intramolecular Friedel-Crafts (Berliner 1949) acetylation.

Smith and collaborators (Smith III et al., 1985) have reported a very interesting rearrangement of the alcohol with sulfuric acid during the synthesis of quadrone, a biologically active sesquiterpene with a unique skeleton.

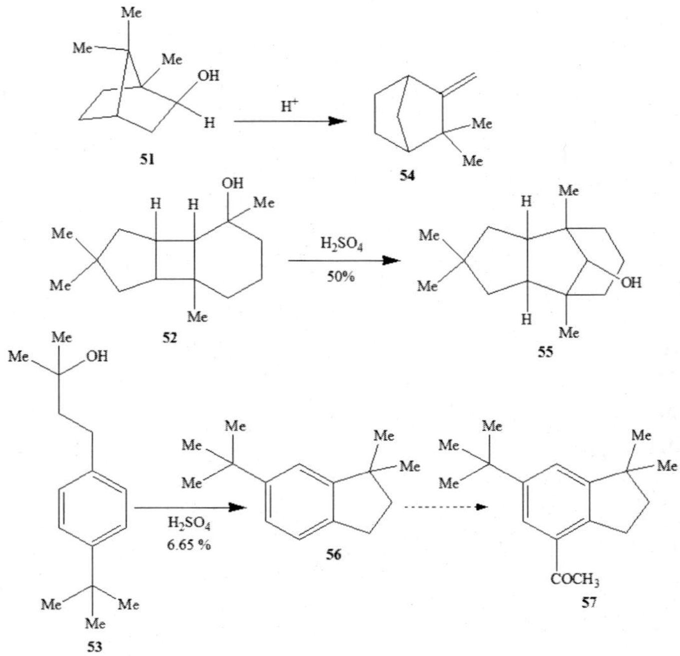

Scheme 9. Rearrangement of alcohols **51, 52** and **53** to isoborneol **54**, α-caryophyllene alcohol **55,** indane 5**6** respectively.

Rearrangement of Diols

Many diols exhibit unexpected rearrangement with sulfuric acid. The reaction of vicinal diol with acid which results in dehydration and migration of an alkyl or aryl group atom to form an aldehyde or ketone is known as Pinacolone (Rickborn 1991) rearrangement which can be viewed as a special case of the Wagneer-Meerwein rearrangement. Many modifications have been introduced over the years for particular glycols, but sulfuric acid remains the most commonly employed. The use of 25% sulfuric acid is recommended in the procedure of Adams (Adams and Adams 1925) to obtain pinacolone in essentially quantitative yield. The pinacol rearrangement reaction can be a useful alternative to the standard methods for the synthesis of aldehydes and ketones. The diol **58** obtained by the dimerization of cyclobutanone smoothly rearranges to ketone **59**, which on warming with sulfuric acid, yields the pentalene **60** (Vogel 1952). On similar treatment, the diol 61 affords (Christol,

Krapcho and Pietrasanta 1969) the ketone 62 and ketone **63** in yields 60% and 40% respectively as is depicted in **Scheme 10**.

The ketone **63** can be converted to the ketone **62** if the temperature is increased (Mundy and Srinivasa 1979). The formation of only the ketone **65** (Mundy, Kim and Warnet 1983) is reported on treatment of the diol **64** with cold concentrated sulfuric acid as exhibited in **Scheme 11**. The transformation occurs via the intermediate **64 (i).**

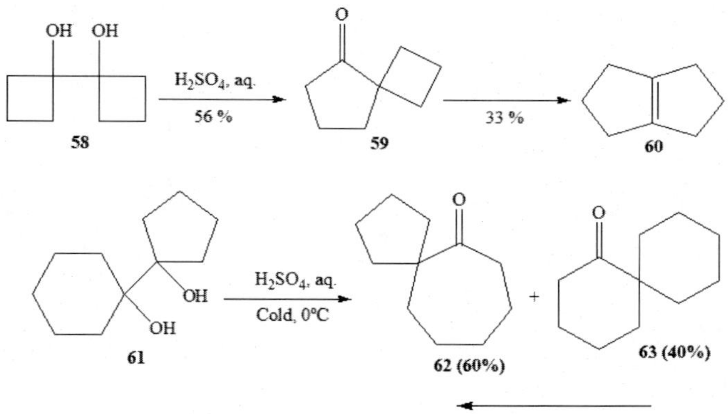

Scheme 10. The Pinacol rearrangement of diols **58** and **61** to pentalene **60** and ketone **62** and **63** respectively.

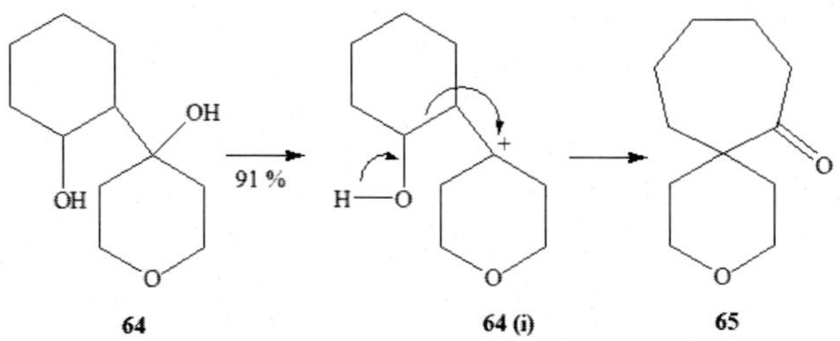

Scheme 11. The Pinacol rearrangement of diol **64** to ketone **65**.

Rearrangement of Acetylenic Alcohols

Molecular rearrangement of acetylenic alcohols with sulfuric acid has been observed. The acid-catalysed isomerization of secondary and tertiary acetylenic alcohols via a [1.3-] shift of the hydroxyl group to the corresponding α,β-unsaturated aldehydes or ketones is known as the Meyer-

Schuster rearrangement (Meyer and Schuster 1922; Smissman et al., 1956; Mundy, Ellerd and Favalero 2005).

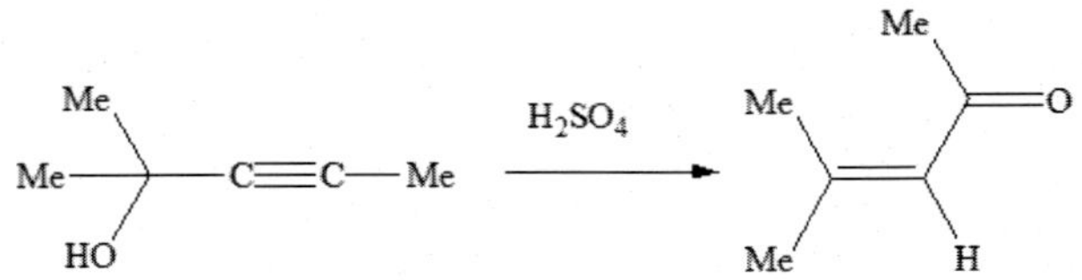

Scheme 12. Rearrangement of acetylenic alcohol with acid.

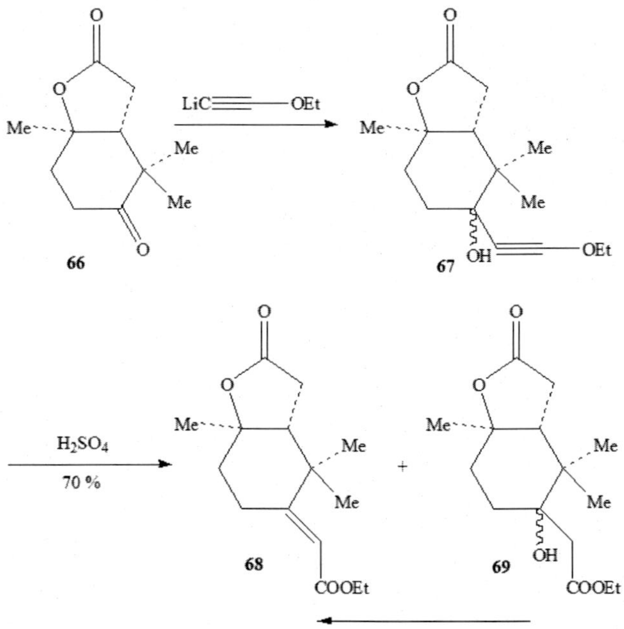

Scheme 13. Rearrangement of alcohol **67** to unsaturated ester **68** by the Meyer-Schuster rearrangement.

The reaction may proceed through an allenic carbonium-ion intermediate from which the nucleophilic attack by water ends up with the formation of α,β-unsaturated carbonyl compound or the reaction may take place (Edens et al., 1977; Andres et al., 1988) through protonation of the hydroxyl group, 1.3-shift of the protonated hydroxyl group and enol-keto tautomerization.

The Meyer-Schuster rearrangement has been utilized for the synthesis of many unsaturated ketones. Thus the propargyl alcohol **67**, prepared from the ketone **66**, on treatment with sulfuric acid affords (Crich, Natarajan and Crich

1997) α,β-unsaturated ester **68** (**Scheme 13**) in good yield along with a little amount of hydroxyester 69 which is smoothly converted to ester 68 with Martin sulfurane reagent PbSoc(CF_3) Ph_2. The unsaturated ester 68 has been synthesized for the construction of the A/B ring of Taxol.

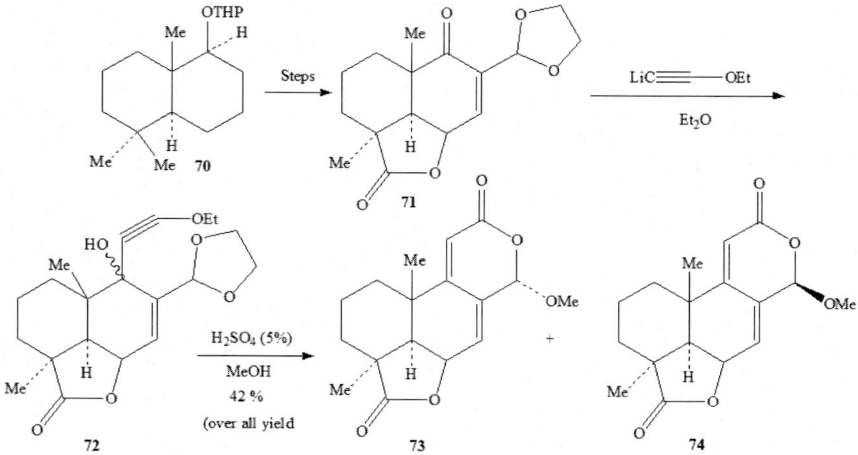

Scheme 14. Synthesis of unsaturated lactone **73** by the Meyer-Schuster rearrangement of alcohol **72**.

The use of Meyer-Schuster rearrangement has also been observed during the synthesis (Welch et al., 1974) of the antibiotics (±) LL-Z12271α **73** which are isolated from the fermentation products of *Acrestalagmus* species and shows the antifungal activity towards several pathogenic fungi. The ketoketal **71**, prepared from the tetrahydropyranyl derivative **70**, on treatment with lithiumacetylide yields an unstable alcohol **72** which, with sulfuric acid, rearranges to the desired lactone **73** along with anomer **74** (**Scheme 14**).

Aromatic Rearrangement

Alkyl groups attached to the benzene ring migrate from one position to another with sulfuric acid. The migration of the alkyl group has also been observed in polyalkyl benzene during sulfonation and this rearrangement is known as the Jacobsen rearrangement (Smith 1942). Durene **75** on sulfonation yields prehnitenesulfone acid **76** which on desulfonation rearranges (Hart and Janssen 1970) to tetramethyl xylene **77**(**Scheme 15**). 4.6-dichloro-m-xylene

78 rearranges (Koch 1890) to 2.4-dichloro-m-xylene **79** and 6.7-dimethyltetramethyl tetralin **80** (yield not specified) affords (Smith and Lo 1948) to 5.6-dimethyl tetralin 81 when are subjected to Jacobsen condition. One of the most important applications of the Jacobsen rearrangement is the isomerization (Schroeter and Gotzsky 1927) of the octahydxroanthracene 82 to phenanthrene **85**. The anthracene **82** on sulfonation affords sulfonic acid **83** which is converted to sulfonic acid **84** by Jacobsen rearrangement. The desulfonation yields the phenanthrene **85**.

It is important to mention that in a number of cases Friedel -Craft alkylation (Berliner 1949; Olah 1963) causes rearrangement involving changes in the relative position of substituents in the benzene ring. The Lewis acid also influences the isomerization (Nightingle 1939) of polyalkylbenzenes.

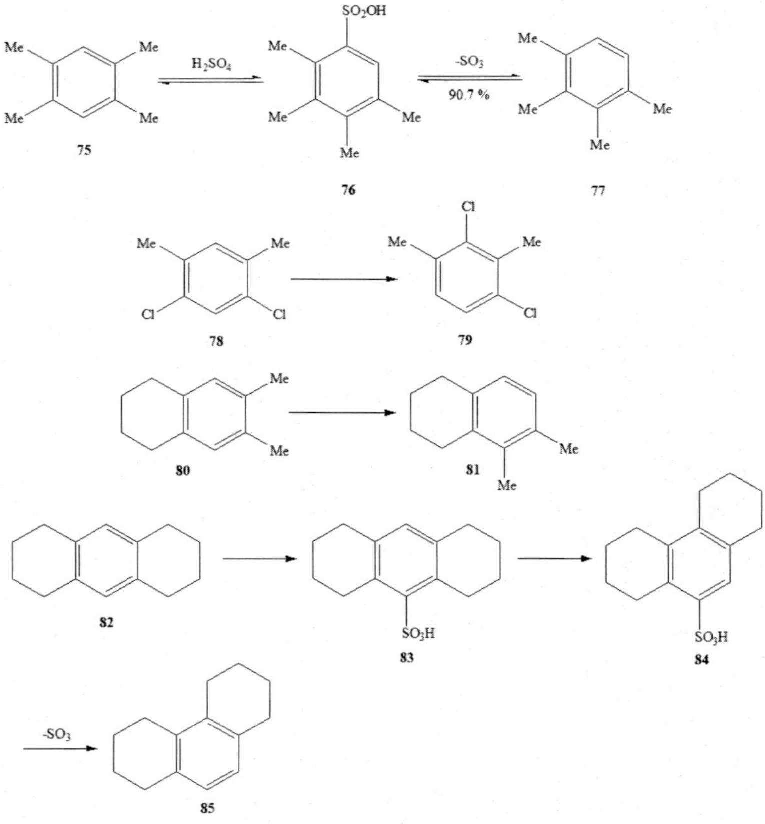

Scheme 15. Synthesis of xylenes 77, 79, tetralin 85 and phenanthrene 88 by Jacobsen rearrangement of xylene 75, 78, 80 and 82.

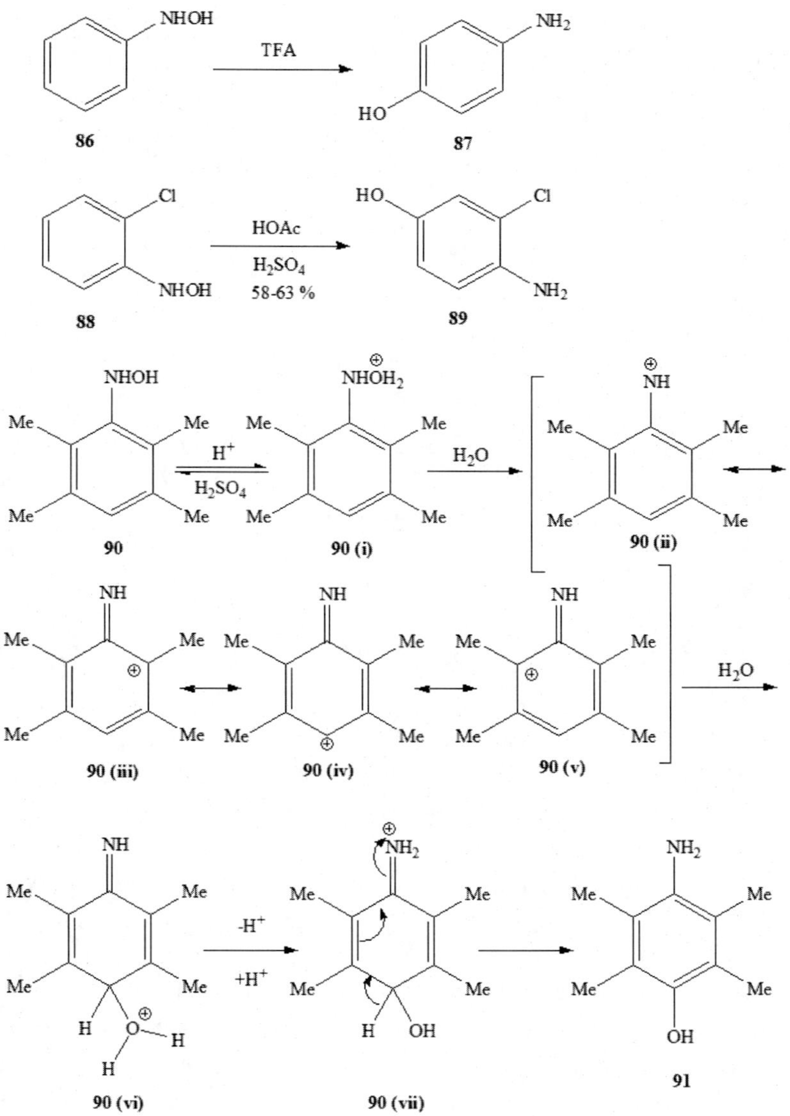

Scheme 16. Synthesis of aminophenols **87, 89** and **91** by the Bamberger rearrangement of phenyl hydroxylamines of **86, 88** and **90** respectively.

Several N-phenylhydroxylamines also suffer rearrangement to aminophenol with sulfuric acid. The rearrangement is known as the Bamberger rearrangement (Bamberger 1894). The N-phenylhydroxylamine **86** undergoes rearrangement with trifluroacetic anhydride (Haga et al., 1992) to yield aminophenol **87** (**Scheme 16**) but its rearrangement with sulfuric acid is not known. The N-phenylhydroxylamine **88** undergoes rearrangement (Harman 1955) with sulfuric acid to 4-aminophenol **89**. The hydroxylamine **90** with sulfuric acid yield (Barcley et al., 1992) amino phenol 91. The mechanism of the transformation of **90** to aminophenol **91** has been shown in Scheme **16.**

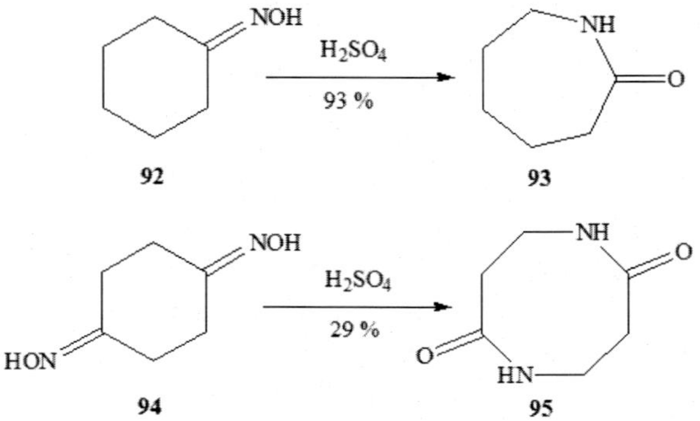

Scheme 17. Rearrangement of oximes **93** and **94** to caprolactams **93** and **95** respectively.

Several oximes also undergo rearrangement with acidic reagents giving N-substituted amides. The rearrangement is known as the Beckmann rearrangement (Laue and Plagens 1999; Gawley 1988), in some cases a mixture of two possible amides may be obtained. The reaction is applied to ketooximes, aldoximes are less reactive. Phosphorous pentachloride is commonly used as a catalyst in the Beckmann rearrangement but sulfuric acid, phosphoric acid, hydrochloric acid, or silicagel can be used. The reaction conditions for the Beckmann rearrangement are often quite drastic (e.g., conc. sulfuric acid at 120ºC) and thus limits the scope to less sensitive compounds. The Beckmann rearrangement of a few oximes with sulfuric acid has been tried but the yield is not very high. The cyclohexanone oxime 92 (Scheme 17) with conc. sulfuric acid yields caprolactam (Eck and Marvel 1943) 93 in 90% yield. The caprolactam 93 is useful for the production of synthetic fibres. The oxime 94 with sulfuric acid yields the lactam (Rothe and Timlar 1962) 95 in 29% yield.

Conclusion and Comments

The present mini review describes the rearrangement of ketones, epoxides, diols, acetylenic alcohols, aromatic compounds, N-phenylhydroxylamine, oximes and other compounds with sulfuric acid. It can be observed that many saturated cyclic ketones undergo rearrangement with sulfuric acid yielding acetophenone and α,β-unsaturated ketone. The Wagneer rearrangement of tertiary alcohol has been utilized to prepare the perfume celestolide. The rearrangement of secondary alcohols with sulfuric acid produces an intermediate that has been utilized for the synthesis of bioactive sesquiterpene quadrone. Many diols with sulfuric acid yield synthetically useful ketones which are difficult to synthesize by the standard methods. Meyer-Schuster rearrangement has proved useful for the synthesis of the A/B ring of Taxol and compounds which show antifungal activity towards several pathogenic fungi. The rearrangement of dienone with sulfuric acid produces phenols. The aromatization of dienone with sulfuric acid and acetic anhydride affords aromatic diacetate. The epoxides with acid can be cleaved to yield ketone. Many tetralones, prepared by the cleavage of tetralene epoxide, have been utilized for the synthesis of sesquiterpeness like occidol, emmotin G methyl ether and bioactive diterpene like carnosic acid. It is interesting to note that alkyl groups attached to the benzene ring migrate on treatment with sulfuric acid. Many products obtained by the rearrangement have not yet received any application in the synthesis of terpenoid compounds but the fundamental skeleton as well as the functional groups attached to these rearrangement products lead us to believe that in near future the compounds can be used by applying suitable organic reactions for the synthesis of terpenoid compounds. The amino phenols can be obtained by the rearrangement of N-hydroxylamine with sulfuric acid. Thus, it is easy to conclude that in the near future many interesting rearrangement of organic compounds with sulfuric acid will be studied.

Acknowledgements

The authors gratefully acknowledge the interest and collaboration of Dr. Edgar Ocando, formerly Chief of the Chemistry Center, IVIC during the preparation of the manuscript.

The collaboration of Nova Science Publishers, Inc. is greatly acknowledged by the authors.

References

Adams, R. and Adams, E. W. 1925. Pinacol hydrate. *Org Syn.* 5: 87-89.

Andres, J., Cardenes, R., Silla, E. and Tapia, O. 1988. A theoritical study of the Meyer- Schuster reaction mechanism: minimum energy profile and properties of the transition-state structure. *J Am Chem Soc.* 110:666-674.

Armstrong, H. E. and Kipping, F. S. 1893. Studies of terpenes and allied compounds. The formation of ketones by interaction of camphor and agents such as sulpuric acid and zinc chloride. *J Chem Soc. Transaction* 63: 75-99.

Arnold, R. T., Buckley, J. S. and Richter, J. 1947. The dienone phenol rearrangement. *J Am Chem. Soc*. 69: 2322-2325.

Ballini, R., Marcantoni, E. and Petrini, M. 1991. An improved and simple synthesis of methyl or ethyl 7-oxoheptanoate and 7-acetoxyheptanal. *Synth Commun.* 21: 1075-1081.

Bamberger, E. 1894. Über das phenylhydoxylamin. *Chem. Ber.* 27: 1548-1557. [*About the phenylhydoxylamine*]

Banerjee, A. K. and Vera, W. J. 2004. Wittig reaction of 5.8-dimethyl-2-tetralone. A formal total synthesis of emotin G-methylether. *J Chem Res (S).* 135-136.

Banerjee, A. K., Carrasco, M. C. and Peña, C. A. 1990. Sulfuric acid-acetic anhydride promoted rearrangement of α,β-unsaturated ketones. *Tetrahedron.* 46: 4133-4136.

Barcley, G. G. Candin, J. P., Lawrie, W., Pulson, P. L. 1992. Amino, aminophrnoxy and aminobenzoyl derivatives of durene. *J. Chem. Res (S).* 245.

Berliner, E. 1949 The Friedel Crafts reaction with dibasic acid anhydride. *Org Reacts*. 5: 229-289.

Cabrera, E. V., Gil, A. C., Ortega, J. G., Bedoya, L., Sanchez, J. and Banerjee, A. K. 2014. A one-pot synthesis of 6-isopropyl-7-methoxy-2-tetralone. *Org Prep Proc Int.* 46: 272-276.

Cabrera, E. V., Sanchez, J. L. and Banerjee, A. K. 2011. A practical synthesis of 7.8-dimethoxy-2-tetralone. *Org Prep Proc Int*. 43: 364-367.

Canonica, S., Ferrari, M. and Sisti, M. 1989. A convenient preparation of ethyl-7-bromoheptanoate. *Org Prep Proc Int.* 21: 253-254.

Cardwell, H. M., Cornforth, F. W., Duff, S. R., Hottermann, H. and Robinson, R. 1951. Total synthesis of androgenic hormone. *Chem Ind*. 389-390.

Christol, H. Krapcho, A. P. and Pietrasanta, F. 1969. Deshydration of pinacols cycliques. *Bull Soc Chim France*. 4059-4062.

Corey, E. J. and Nozoe, S. 1965. The total synthesis of α-caryophyllene alcohol. *J Am Chem Soc.* 87: 5733-5735.

Crich, D., Natarajan, S. and Crich, J. Z. 1997. Synthesis of the taxol AB-system by olefination of an A ring C1 ketone direct B-ring closure. *Tetrahedron*. 53: 7139-7158.

De Mayo, P. 1964. *Molecular Rearrangements,* Vol 2, 709-1239. Interscience Publishers.

Eck, J. C. and Marvel, C. S. 1943. 2-ketohexamethylenime. *Coll. Org. Syn.* Vol 2. 371-372.

Edens, M., Boerner, D., Chase, C. R. Nass, D. and Schiavelli, M. D. 1977 Mechanism of the Meyer-Scuster rearrangement. *J Org Chem.* 42:3403-3408.

Elison, R. A. 1973. Methods for the synthesis of 3-oxocyclopentenes. *Synthesis.* 7: 397-412.

Gawley, R. E. 1988. The Beckmann Reactions: rearrangements, elimination-additions, fragmentations and rearrangement-cyclizations. *Org React*. 35: 1-420.

Graham, S. L., Heathcock, C. H., Pirrung, M. C., Plavac, F. and White, C. T. 1983. Total Synthesis of Sesquiterpenes (1970-1979) in *Total Synthesis of Natural Products*, vol 5, edited by Apsimon, J., John-Wiley Sons, N. Y. p. 1-541.

Haga, N., Endo, Y., Kataoka, K. I., Yamaguchi, K., Shudo, K. 1992. Acid catalyzed amine-migration of O-phenylhydroxylamines. *J Am Chem Soc.* 114: 9795-9806.

Harman, R. E. 1955. Chloro *p*-benzoquinone. *Org Syn.* 35: 148-154.

Hart, H. and Janssen, J. F. 1970. Jacobsen-type rearrangements in aromatic trichloromethylation. *J Org Chem.* 35: 3637-3641.

Hayakawa, T., Nakamura, H., Aoki, K., Suzuki, M., Yamada, K. and Hirata, Y. 1971. Synthetic intermediate potentially useful for the synthesis of dendobrine. *Tetrahedron*. 27: 5157-5162.

Heathcock, C. H. 1973. Total Synthesis of Sesquiterpenes. In *Total Synthesis of Natural Products*, vol 2, edited by Apsimon, J. John-Wiley & Sons, New York. 2, p. 197-558.

Hogeween, H. and Vankruchen, E. M. G. A. 1979. Wagner-Meerwein rearrangements in long-livedpolymethyl substituted bicycle [3.2.0] heptadienyl cations. *Topics in Curr. Chem.* 80: 90-124.

Kagabu, S. and Kojima, Y. 1972. A synthesis of indane musk celestolide. *J Chem Edu*. 84:420.

Kato, M., Yamamoto, S., Nomura, S. and Miwa, T. 1990. One pot synthesis of substituted tropanes from 7.7-dihalo-2.3-(or 3.4)-epoxybicyclo [4.1.0] heptane derivatives. *Bull Chem Soc. Jpn*. 63: 64-73.

King, J. F. and de Mayo, P. 1964. *Terpenoid Rearrangements in Molecular Rearrangements*, vol 2. Edited by de Mayo, P. Interscience Publications, New York, p 771-840

Koch, E. 1890. Über Dibrom-und dichlorxylole, speciell die durch schweflsäure bewrikte Umlagerung der symmtrischen verbindungen. *Chem Ber*. 23:2318-2321.

Kropp, P. J. 1963. On the mechanism of the dienone-phenol rearrangement of 2-methyl methyl-1.4-dien-3-one. *Tetrahedron Lett*. 1671-1676.

Laue, T., Plagens, A. 1999. *Named Organic Reactions*. John-Wiley & Sons., NY. p 22-24.

Levine, S. G. 1960. The structure of isocamphorquinone. *J Am Chem Soc*. 82: 25562559.

Lutz, R. P. and Roberts, J. D. (1962). The mechanism of the rearrangement of fenchone. *J. Am. Chem. Soc.* 84: 3715-3721.

Marsh, J. E. 1899. The action of sulphuric acid on fenchone. *J Chem Soc*. 75: 1058-1060.

Marvell, E. N. and Geiszler, A. 1952. The rearrangement of some substituted cyclic ketones: an usual dienone-phenol rearrangement. *J Am Chem Soc.* 74: 1259-1263.

Marvell, E. N. and Stephenson, J. L. 1955. The dienone–phenol rearrangement. A new example of abnormal path. *J Am Chem Soc.* 77: 5177-5179.

Mclamore, W. M. 1951. Synthesis of some hydroxyl alkylbenzquinones. *J. Am. Chem. Soc.* 73: 2225-2230.

Meyer, K. H. and Schuster, K. 1922. Unlagerrung tertiärer Äthinyl-carbinole in ungesäattigte ketone. *Chem. Ber*. 55B: 819-823.

Mundy, B. P. and Srinivasa, R. 1979. Ring size effects. II. An examination of conditions for the pinacol rearrangement. *Tetrahedron Lett.* 29: 2671-2672.

Mundy, B. P. Ellerd, M. G. and Favalero, F. G. 2005. *Name Reactions and Reagents for Organic Synthesis,* John -Wiley, New York, p. 426-427.

Mundy, B. P., Kim, Y. and Warnet, R. 1983. The synthesis of spiro (5.6)-3-oxaundecan-7-one, observations of a specific oxygen-directing effect. *Heterocycles*. 20: 1727-1730.

Nightingle, D. V. 1939. Alkylation and the action of aluminiumhalides on alkylbenzenes. *Chem. Reviews*. 25: 329-376.

Noyce, D. S. 1950. A rearrangement of camphenilone. *J Am Chem Soc.* 72: 924-925.

Olah, G. A. 1963-1964. *Friedel-Crafts and Related Reactions.* Edited by Olah, G. A. Wiley. New York Vol 1 and Vol 2. P. 1-1930.

Pirrung, M. C. Morehead Jr., A. T., and Young, B. G. 2000. Bicyclic Sesquiterpenes. In the *Total Synthesis of Natural Products,* vol 11., edited by Apsimon, J., John-Wiley, New York, p 1-455.

Rickborn, B. 1991. Pinacol hydrate. In *Comprehensive in Organic Synthesis,* vol 3, edited by Trost, B. and Flemming, I., Pergamon Press, Oxford, p 721-732.

Rothe, M., and Timlar, R. 1962. Beckmann und Schmidt Umlagerung alicyclisehr Diketone. Synthese von cyclodiamiden des mitteren Ringberiches. *Chem. Ber.* 95: 783-794. [Beckmann and Schmidt rearrangement alicyclic very diketones. Synthesis of cyclodiamides of the middle ring region]

Schroeter, G. and Gotzsky, S. 1927. Über hydrierung des anthracensz Mitteilung: unwandlungen der okthacen-sulfonsäure. *Chem Ber.* 60: 2035-2045.

Smissman, E. E., Johnsen, R. H., Carlson, A. W., and Aycock, B. F. 1956. The acid catalyzed rearrangement of phenylethylnyl carbinols. *J Am Chem Soc.* 78: 3395-3400.

Smith III, A. B., Wexler, B. A., Tu, C. Y. and Konopelski, J. P. 1985. Stereoelectronic effects in the cationic rearrangements of [4.3.2] propellenes. *J Am Chem Soc.* 107: 1308-1320.

Smith, L. I. 1942. The Jacobsen reaction. *Org React.* 1: 370-374.

Smith, L. I. and Lo, C. P. 1948. The Jacobsen reaction. 6.7-dialkyltetralins. *J Am Chem Soc.* 70: 2209-2215.

Soffer, M. D., Stewart, R. A., Cavagnal, J. C., Gellerson, and H. E. and Bowler, E. 1950. Synthesis in the direction of morphine. Some intermediates and model compounds. *J Am Chem Soc.* 72: 3704-3709.

Sorensen, T. S. 1976. Terpene rearrangements from superacid activities. *Account Chem Res.* 9: 257-265.

Streitweiser, Jr, A. 1956. Solvolytic displacement reactions at saturated carbon atoms. *Chem Rev.* 56: 571-752.

Tahara, A., Miizuno, H., and Ohsawa, T. 1972. Methyl migration in dehydroabietic acid derivative. *Chem Lett.* 1163-1165.

Thompson, R. H. 1992. Total Synthesis of Naturally occurring quinones. In *Total Synthesis of Natural Products*, vol 8, edited by Ap Simon, J., John-Wiley, NY. P 197-558.

Thompson, S. K. and Heathcock, C. H. 1992. Total synthesis of some marasmane and lactaranes. *J Org Chem.* 57: 5979-5989.

Vera, W. J. and Banerjee, A. K. 2006. A formal total synthesis of (±)occidol. *Syn Commun.* 36: 3091-3094.

Vogel, E. 1952. Eine synthesie del 7.8-dehydro-pentalans[Bicyclo. 3.3] octens-(7,8). *Chem Ber.* 85: 25-29.

Welch, S. C., Hagan, C. P., White, D. H., and Fleming, W. P. 1974. A total synthesis of antifungal mold metabolite (±) LL-Z1271α. *Syn Commun.* 4:373-378.

Woodward, R. B and Singh, T. 1950. Synthesis and rearrangement of cyclohexadienone. *J Am Chem Soc.* 72:494-500.

Chapter 6

Alkoxylation of Terpenes over Heterogenous Catalysts

J. E. Castanheiro[1,*], A. P. Pinto[1] and M. E. Lopes[2]
[1]MED, Departamento de Química, Escola de Ciências e Tecnologia, Universidade de Évora, Évora, Portugal
[2]Hercules, Departamento de Química, Escola de Ciências e Tecnologia, Universidade de Évora, Évora, Portugal

Abstract

Monoterpenes are renewable raw material. They can be used in the pharmaceutical, cosmetic and food industries as active components of drugs and ingredients of fragrances. Alkoxylation reactions of terpenes are important routes to obtain compounds with high commercial value. Traditionally, homogeneous acid catalysts (sulfuric acid) have been used in these reactions. However, these homogenous catalysts have some disadvantages such as, its separation from reaction mixture, they cannot be reused. The solid acids have many advantages over liquid acids. They do less harm to the environment and do not have corrosion or effluent disposal problems. They are reusable and easy to be separated from liquid products. This review focuses on alkoxylation of terpenes over heterogeneous catalysts.

Keywords: limonene, pinene, camphene, heterogenous catalysts

---

* Corresponding Author's E-mail: jefc@uevora.pt.

In: Terpenes and Terpenoids
Editor: Charles A. Davies
ISBN: 978-1-68507-559-0

Introduction

Monoterpenes are renewable raw materials used for the synthesis of compunds with high commercial valeu, which can be used as flavour and fragrances in fine chemistry [1-4]. Different catalytic reactions, like isomerization, hydration and acetoxylation of terpene have been performed using heterogeneous catalysts.

The alkoxylation of alpha-pinene, beta-pinene, limonene and camphene yield a mixture of terpenic ethers which exhibit organoleptic properties, and thus, they are used in various fragrance compositions. Also, terpenic ethers exhibit organoleptic properties [1-4]. Traddicionally, these reactions have been carried out over homogenous catalysts. However, environment problems, the homogenous processes have been replaced by heterogeneous ones. The use of heterogeneous catalysts have some advantages such as easy separation from reaction mixture, possibility of reutillization and regeneration.

In this work, will be made a review of terpenes alkoxylation, hydration, esterification over solid acid catalysts.

The Figure 1 shows the scheme of alkoxylation of limonene with an alcohol over an acid catalyst.

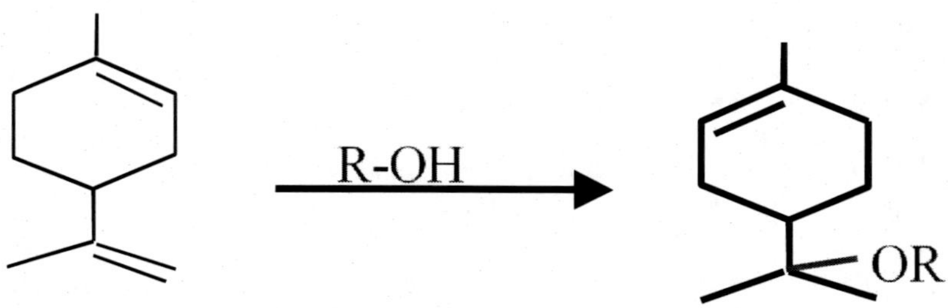

Figure 1. Scheme of reaction between limonene with an alcohol.

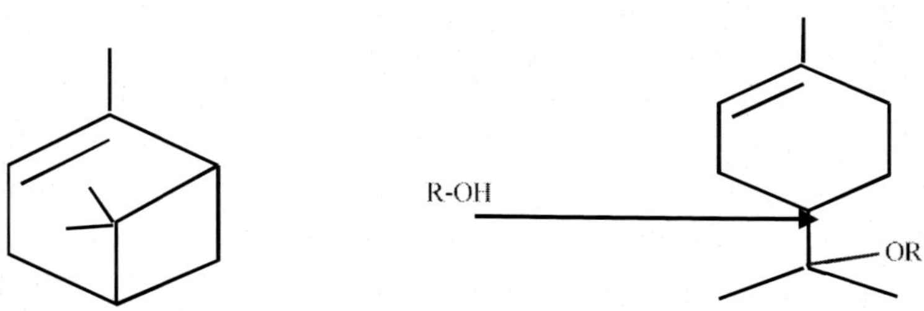

Figure 2. Reaction scheme of α-pinene with an alcohol.

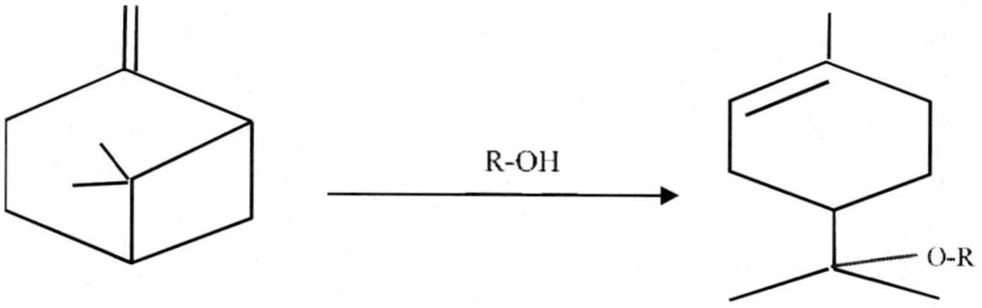

Figure 3. Reaction scheme of β-pinene with an alcohol.

Another rection that have been studied, it is the alkoxylation of a-pinene with an alcohol (methanol, ethanol, propan-1-ol and butan-1-ol) over solid catalysts (Figure 2).

The alkoxylation of b-pinene with different alcohols has been studied over several heterogeneous catalysts, Figure 3 shows the reaction of the b-pinene alkoxylation.

The reaction between camphene with an alcohol have been studied over heterogeneous catalysts. Figure 4 shows the camphene alkoxylation.

Zeolites

Zeolites are aluminosilicates crystalline formed by the AlO_4^- or SiO_4 tetrahedra. These materials show shape-selectivity, acid-base properties, high surface area, uniform microporosity and hydrophobicity. Zeolites have been used as heterogeneous catalyst in petrochemical industry and fine chemical processes, due to their strong acidity, excelent stability and regular pore sizes [5-10].

Beta zeolite was used as heterogenous catalysts in alkoxylation of a-pinene and limonene with Cl-C5 alcohols to 1-methyl-4-[a-alkoxy-isopropyl]-l-cyclohexene using a batch and a continuous fixed bed reactor [11]. In both reactors, the use of methanol as an addition compound and limonene as feedstock gives 1-methyl-4-[alpha-methoxy-isopropyl]-l-cyclohexene with the yield of 85% (conversion: 93%, selectivity: 92%). By means of variation of the reaction parameters, the limonene conversion can be adjusted within the range 40-90%. The selectivity to 1-methyl-4-[a-methoxy-isopropyl]-l-cyclohexene remains always about 95%. For mixtures of methanol and alpha-pinene, the highest yield of corresponding a-terpinyl methyl ether is only of about 50% (conversion: 99%, selectivity: 51%) in both reactor types, as several bicyclic and double addition products are formed in parallel. The selectivity of all addition products can reach values of about 85%. In the case of limonene and alpha-pinene alkoxylation with higher alcohols (C2-C5 alcohols), good results are obtained by using linear alcohols only. Other (iso- and branched) alcohols tend to dehydrate to corresponding alkenes and water. This might also reflect a shape selective effect. Alkoxylation of a-pinene or limonene in the presence of other large pore zeolites, e.g., H-USY or H-mordenite, and solid acids results in a considerable selectivity reduction to the corresponding a-terpinyl alkyl ethers. The destruction of the structure of beta zeolite leads to a complete deactivation of the catalyst [11].

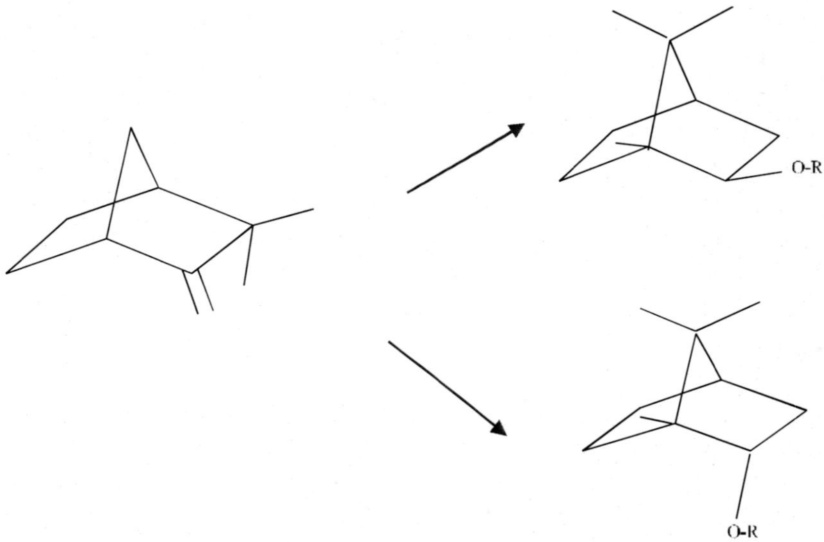

Figure 4. Reaction scheme of camphene alkoxylation.

Mesostructured Silica

Mesostructured silica materials, such as MCM-41, PMOs and SBA-15 have been used in heterogeneous catalysis as catalyst supports, due to a combination of high surface areas and controlled pore sizes. The functionalization of organic groups onto the surface of these materials could be by grafting on the surface or by co-condensation during the synthesis. Many silylation agents have been used to functionalise the surface of the mesoporous materials, such as 3-aminopropyltrimethoxysilane and 3-mercaptopropyltrimethoxysilane. These modified mesoporous silicas have been used as catalyst in a wide range of chemical reactions, depending on the nature of the functional group. Direct method usually yields a uniform surface coverage with organic groups and provides better control over the number of organic groups incorporated to the structure. However, products obtained by post-synthesis grafting are often structurally better defined and hydrolytically more stable than samples obtained from direct synthesis method [12].

Sulfonic acid-functionalized mesostructured silica (MCM-41, PMO) have been used as catalyst in the methoxylation of α-pinene at 60°C [13]. The support functionalization was achieved by the introduction of 3-(mercaptopropyl)trimethoxysilane onto the surface of these materials either by grafting or by co-condensation. The thiol groups were oxidized to SO_3H by treatment with H_2O_2. All the catalysts were active in the studied reaction, PMO-SO_3H-g being the best one. Good values of selectivity to alpha-terpinyl methyl ether were obtained with these catalysts. Catalytic stability of the PMO-SO_3H-g was evaluated by performing consecutive batch runs with the same catalyst sample. After the third batch, a stabilization of the activity was observed [13].

Heteropolyacids

Catalysis by heteropoly acids (HPAs) and related polyoxometalate compounds is a field of increasing importance. HPAs have several advantages as catalysts which make them economically and environmentally attractive. On the one hand, HPAs have a very strong, approaching the superacid region, Bronsted acidity; on the other, they are efficient oxidants, exhibiting fast reversible multielectron redox transformations under rather mild conditions. Their acid-base and redox properties can be varied over a wide range by changing the chemical composition. Solid HPAs possess a discrete ionic structure, comprising mobile basic structural units heteropolyanions and countercations (H^+, H_3O^+, $H_5O_2^+$, etc.) sunlike the network structure of, e.g., zeolites and metal oxides [14].

$H_3PW_{12}O_{40}$, $H_3PMo_{12}O_{40}$, and $H_4SiW_{12}O_{40}$ and $H_4SiMo_{12}O_{40}$ immobilized on silica were used in the α-pinene methoxylation [15]. The main product of α-pinene methoxylation was α-terpinyl methyl ether. It was observed that the catalytic activity increases in the series: SiMo < PMo_S < SiW_S < PW2_S. Catalysts consisting of the PW immobilized on silica with different PW loading were prepared. It was observed that the catalytic activity increases with the amount of PW immobilized on silica. However, at a high amount of PW on silica, a decrease of the catalytic activity was observed. All the catalysts tested on α-pinene methoxylation exhibited good selectivity (about 60% at near complete alpha-pinene conversion). The catalytic stability of the PW2_S was evaluated by performing consecutives batch runs with the same catalyst sample. Similar activity for all the catalytic tests was observed [15].

The acid-catalyzed alkoxylation of α-pinene, β-pinene and limonene with Cl-C4 alcohols (methanol, ethanol, 1-propanol and 1-buthanol) to α-terpinyl alkyl ether was carried out in the presence of silica-occluded $H_3PW_{12}O_{40}$ [16]. For the alkoxylation of α-pinene, β-pinene and limonene, the catalytic activity decreases with an increase in the chain length of the linear alcohols. This behavior could be explained due to the presence of sterical hinderance and diffusion restrictions in the porous system of the catalyst. The selectivity of silica-occluded heteropolyacid to α-terpinyl alkyl ether also decreased with the increase of the chain length of the alcohols. The catalytic stability of the PW-S was studied in the ethoxylation of α-pinene and β-pinene. It was observed that the catalyst can easily be recovered and reused without loss of activity. It can be also observed that the alkoxylation reactions with beta-pinene substrate were faster than the α-pinene and limonene [16].

SBA-15-occluded $H_3PW_{12}O_{40}$ was used as catalyst on the alkoxylation of α-pinene, β-pinene and limonene [17]. A series of catalysts with different $H_3PW_{12}O_{40}$ amount varying from 1.8 to 19.3 wt.% were prepared. PW4-SBA-15 (with 10.8 wt.%) exhibited the highest catalytic activity for the alkoxylation of α-pinene with ethanol. An approximately 53% selectivity to α-terpinyl ethyl ether was observed over the PW-SBA-15 catalysts. PW4-SBA-15 was also used as a catalyst for the alkoxylation of other terpenes, including -pinene and limonene. The catalytic activity increased in the following order: limonene < α-pinene < β-pinene. The PW4-SBA-15 catalyst exhibited high catalytic stability for the alkoxylation of alpha-pinene [17].

The alkoxylation of camphene with C1-C4 alcohols (methanol, ethanol, 1-propanol, 2-propanol, 1-butanol, and 2-butanol) to alkyl isobornyl ether was successfully carried out in the presence of silica-occluded $H_3PW_{12}O_{40}$ [18]. The catalytic activity decreases with an increase of the chain length of the linear alcohols, which can be explained due to the presence of diffusion restrictions in the porous system of catalyst and sterical hindrance. High selectivity for alkyl isobornyl ether was obtained in the presence of silica-occluded heteropolyacid. The catalyst (PW-Ssg) can be recovered and reused without loss of activity and selectivity. The stability of PW-Ssg catalyst towards metal leaching (less than 1%) is much greater than that of PW-Sim which lost 20% of its initial amount of heteropolyacid, during the catalytic reaction [18].

The alkoxylation of camphene with methanol, ethanol and 1-butanol was studied in the presence of $H_3PW_{12}O_{40}$ and $Cs_{2.5}H_{0.5}PW_{12}O_{40}$ [13]. In the same work, de Meireles et al. [19] studied the alkoxylation of α-pinene, β-pinene and limonene with 1-butanol over $H_3PW_{12}O_{40}$ and $Cs_{2.5}H_{0.5}PW_{12}O_{40}$. The authors observed that the heteropolyacid $H_3PW_{12}O_{40}$ and its acidic Cs salt, $H_3PW_{12}O_{40}$, are efficient catalysts for the liquid phase alkoxylation of camphene, limonene, α-pinene, and β-pinene. These reactions provide effective routes for the synthesis of isobornyl and α-terpenyl ethers [19].

Activated Carbons

The methoxylation of α-pinene was performed using activated carbons (a commercial microporous carbon and a xerogel mesoporous carbon) as heterogeneous catalysts [20]. It was observed that the catalysts prepared from the biomass waste used (olive stones), it can be possible to obtain a microporous carbon which can later be used as an acid catalyst, resulting in a clear added value to the material. The catalytic activity increased in the order Cnorit< CMN< CB. The results also seem to indicate that the most important feature of these catalysts is the surface chemistry, although some effect of the textural properties could be observed. All catalysts showed good selectivity towards the α-terpinyl methyl ether. These results show that it is possible to reutilize a waste residue and convert it to high added value product such as an acid catalyst [20].

Other Catalyst

Methoxylation of α-pinene was carried out over poly(vinyl alcohol) (PVA) containing sulfonic acid groups have been used as catalyst in [21]. These groups were introduced by the direct reaction between –OH groups of PVA and sulfosuccinic acid (SSA). PVA catalysts were prepared using sulfosuccinic acid (SSA) by adjusting the amount of SSA (5-40 mol%). It was observed that the amount of acid sites increases with the amount of SSA. It was also observed that the diffusivity of alpha-pinene through the polymeric catalysts decreases with the increase of the crosslinking degree. The conversion of alpha-pinene increases when the amount of sulfosuccinic acid used in the polymer crosslinking is increased from 5% to 40%. However, when the crosslinking degree increases from 20% to 40%, the conversion of alpha-pinene increases only slightly. Good values of selectivity to α-terpinyl methyl ether (about 60% near complete conversion) were obtained over PVA with sulfonic acid groups. Catalytic stability of the PVA_SSA20 was evaluated by performing consecutive batch runs with the same catalyst sample. After the third batch, a stabilization of the initial activity was observed [21].

Conclusion

Monoterpenes are natural compounds present on biomass, as an essential oil. These compounds are an important biomass-based feedstock for the fragrance and pharmaceutical industries. Traditionally, the valorization of terpenes into a compound with high commercial value is performed using homogenous catalysts, which have some disadvantages, such as production of toxic waste and its difficulty in separation from reaction mixture. To convert these different processes into "green process," these reactions have been carried out over different heterogeneous catalysts. The use of heterogeneous catalysts for terpenes valorization has been applied in different reactions.

References

[1] Mäki-Arvela, P; Holmbom, B; Salmi, T; Murzin, DY. 2007. "Recent Progress in Synthesis of Fine and Specialty Chemicals from Wood and Other Biomass by Heterogeneous Catalytic Processes." *Catalysis Reviews* 49:197-340.

[2] Gusevskaya, EV. 2014. "Reactions of Terpenes Catalyzed by Heteropoly Compounds: Valorization of Biorenewables." *ChemCatChem* 6:1506-1515.

[3] Corma, A; Iborra, S; Velty, A. 2007. "Chemical Routes for the Transformation of Biomass into Chemicals," *Chemical Reiews* 107:2411-2502.

[4] Gupta, P; Paul, S. 2014. "Solid acids: Green alternatives for acid catalysis." *Catalysis Today* 236:153-170.

[5] Cundy, CS; Cox, PA. 2003. "The hydrothermal synthesis of zeolites: history and development from the earliest days to the present time." *Chemical Reviews* 103: 663-701.
[6] Corma, A. 1995. "Inorganic solid acids and their use in acid-catalyzed hydrocarbon reactions." *Chemical Reviews* 95:559-614.
[7] Anis, SF; Khalil, A; Saepurahman, SG; Hashaikeh, R. 2016. "A review on the fabrication of zeolite and mesoporous inorganic nanofibers formation for catalytic applications." *Microporous and Mesoporous Materials* 236:176-192.
[8] Davis, ME; Lobo, RF. 1992. "Zeolite and molecular-sieve synthesis." *Chemical Materials* 4:756-768.
[9] Davis, ME. 2014. "Zeolites from a materials chemistry perspective." *Chemical Materials* 26:239-245.
[10] Corma, A. 1997. "From microporous to mesoporous molecular sieve materials and their use in catalysis." *Chemical Reviews* 97:2373-2419.
[11] Hensen, K; Mahaim, C; Hölderich, WF. 1997. "Alkoxylation of Limonene and Alpha-Pinene over Beta-Zeolite as Heterogeneous Catalyst." *Applied Catalysis A: General*, 149:311-329.
[12] Melero, JA; van Grieken, R; Morales, G. 2006. "Advances in the Synthesis and Catalytic Applications of Organosulfonic-Functionalized Mesostructured Materials." *Chemical Reviews* 106:3790-3812.
[13] Castanheiro, JE; Guerreiro, L; Ramos, AM; Fonseca, IM; Vital, J. 2008." Mesoporous silica containing sulfonic acid groups as catalysts for the alpha-pinene methoxylation." *Studies in Surface Science and Catalysis*, 174:1319-1322.
[14] Kozhevnikov, IV. 1998. "Catalysis by Heteropoly Acids and Multicomponent Polyoxometalates in Liquid-Phase Reactions." *Chemical Reviews* 98:171-198.
[15] Pito, DC; Matos, I; Fonseca, IM; Ramos, AM; Vital, J; Castanheiro, JE. 2010. "Methoxylation of α-pinene over heteropolyacids immobilized in silica." *Applied Catalysis A: General*, 373:140-146.
[16] Caiado, M; Pito, DS; Castanheiro, JE. 2013. "Alkoxylation of Terpenes over Tungstophosphoric Acid Immobilised on Silica Support." In: *Environmentally Benign Catalysts: For Clean Organic Reactions*, edited by A. Patel, 153-164, Springer.
[17] Castanheiro, JE; Fonseca, IM; Ramos, AM; Vital, J. 2017. "Tungstophosphoric acid immobilised in SBA-15 as an efficient heterogeneous acid catalyst for the conversion of terpenes and free fatty acids." *Microporous Mesoporous Materials* 249:16-24.
[18] Caiado, M; Machado, A; Santos, RN; Matos, I; Fonseca, IM; Ramos, AM; Vital, J; Valente, AA; Castanheiro, JE. 2013. "Alkoxylation of camphene

over silica-occluded tungstophosphoric acid." *Appied Catalysis A: General* 451:36-42.

[19] de Meireles, ALP; Costa, MS; Rocha, KAS; Kozhevnikova, EF; Kozhevnikov, IV; Gusevskaya, EV. 2014. "Heteropoly acid catalysts for the synthesis of fragrance compounds from biorenewables: The alkoxylation of monoterpenes." *Chem Cat Chem* 6:2706-2711.

[20] Matos, I; Silva, MF; Ruiz-Rosas, R; Vital, J; Rodríguez-Mirasol, J; Cordero, T; Castanheiro, JE; Fonseca, IM. 2014. "Methoxylation of α-pinene over mesoporous carbons and microporous carbons: A comparative study." *Microporous and Mesoporous Materials*, 199:66-73.

[21] Pito, DS; Fonseca, IM; Ramos, AM; Vital, J; Castanheiro, JE. 2009. "Methoxylation of α-pinene over poly(vinyl alcohol) containing sulfonic acid groups." *Chemical Engineering Journal,* 147:302-309.

Chapter 7

Characteristics and Applications of Terpenes and Terpenoids from Brazilian Flora

L. A. A. P. Fonseca* **and M. A. Cremasco**
School of Chemical Engineering, University of Campinas,
Campinas, Brazil

Abstract

This chapter presents a discussion about the importance of Brazilian flora for herbal medicine. Ethnomedical knowledge was possessed by different Brazilian native groups and consisted of complex practices, especially with the use of native herbs. Terpene and terpenoid compounds are natural products isolated from different natural sources. It consists of five-carbon isoprene units, which are assembled from many isoprene units. Terpenes are simple hydrocarbons, while terpenoids are modified terpenes with different functional groups (ketones, alcohols, esters, carboxylic acids, aldehydes) and oxidized methyl groups moved or removed at various structural positions. Terpenes and terpenoids are key components of essential oils. Brazil is considered the country with the greatest biodiversity in the world. According to the Brazilian Ministry of Environment, at least 103,870 animal species and 43,020 plant species are known. These species are distributed in six biomes with different characteristics and vegetation: Amazon Rainforest, Cerrado, Atlantic Forest, Caatinga, Pantanal, and Pampa. The wisdom of Brazilian natives

* Corresponding Author's E-mail: lucianapreviato2014@gmail.com.

In: Terpenes and Terpenoids
Editor: Charles A. Davies
ISBN: 978-1-68507-559-0

transcends historical time and extermination to offer hope, especially in times of pandemics.

Keywords: Brazilian flora, essential oil, terpene, terpenoid

Introduction

When the Spaniards arrived in the Americas in the 16th century, the South American countries as we know them today did not exist; however, parts of Uruguay, Paraguay, Argentina, and Brazil were identified as the territory of Guayrá, currently located in the state of Paraná, Brazil. The Guarani people inhabited the jungles, and their culture was considered uncivilized by Europeans in that century as well as today [1]. When Guayrá natives are referred to as crude, rude or uncultured, this elicits reflection on Western civilization's ethnocentric assumption that a society is strengthened insofar as it accumulates and spreads knowledge. This occurs, among other means, through the recording of history, offering the impression that writing and the progress of society are part of a linear and irreversible process. The written or great works are evidence of this process, constituting a tradition. In this tradition, nonliterate societies were considered savages who lacked history. The Guarani are a typical example as their exclusion was decreed in different ways. Initially, for an idea already acquired: they contributed nothing remarkable, no durable product before the arrival of so-called civilization, mainly from Europe.

For this reason, it is essential to emphasize the vast knowledge of the Brazilian natives, originating mainly from manifestations of nature. One of these is Kaa'pora or the forest spirit, which manifested itself in several ways, including in medicinal herbs against snake venom [2] or stomach ailments [3].

The genesis of herbal medicine in Brazil is directly correlated with the cultural identity of the country. Before arrival of the colonizers, there were around 200,000 native nations in Brazil [4]. This cultural richness is still evident in the variety of native languages found among native Brazilians, totaling 188 languages, including Guarani, Arawake-Maipure, Yanomami, and Tupi in addition to dialects such as Xavante and Tapaiúna [5].

The Brazilian natives had a vast knowledge of plants; for example, the Guarani had a more profound comprehension of plants than the Europeans of the 16th century [6, 7]. In the Guarani culture, the domain of medicinal properties associated with botanical species that permeated their daily life was one of the most solid accounts of their botanical and therapeutic knowledge, highly adapted to the contexts in which they lived. Medicinal plants, among which those found mainly in the *Myrtaceae, Fabaceae, Asteraceae, Lauraceae, Poaceae* families [8] are highlighted.

It should be noted that the Kaiowá of the Guarani linguistic trunk and originating from the Kaa'gua, which mean "those of the forest," are traditionally farming people having a strong territorial identity. Guarani-Kaiowá cosmology and knowledge have a sacred dimension of the earth and its inhabitants. Myth reflects their way of being [9]. When someone does not acknowledge the wisdom of natives, they ignore ancient knowledge as of medicinal herbs [10, 11] found in Brazil. In times of pandemic where drugs are sought to combat Covid-19, plants with therapeutic uses related to the respiratory system cannot be overlooked.

Forest peoples, whose wisdom comes from the heart of the jungle, know applications for a variety of medicinal species. In addition to the Fabaceae family (species *Cedrela fissilis* and *Pterodon emarginatu*), the Amaranthaceae (species Moquiniastrum polymorphum and Achyrocline satureioides) and Bignociaceae families (species Ananas ananassoides) can be mentioned [12]. When a people or nation is said to come from the jungle, this is far from an insult. It implies recovery of an immense cultural wealth, the so-called traditional community's worldvie with cultural and symbolic practices.

The Kayapó methods and techniques for handling species in Amazon are considered the path for discovering new foods, medicines, dyes, and essential oils [13]. It is worth emphasizing the importance of native knowledge for Brazilian culture, including for herbal medicine, especially in times when we are experiencing a new genocide of indigenous peoples in Brazil [14]. Approximately five hundred years later, herbal medicine's promising applications increase the interest in studying essential oil compounds such as terpenes and terpenoids for creating an effective and culturally appropriate option [15].

Essential oils are obtained from plant material, including flowers, roots, bark, leaves, seeds, and fruits [16]. Terpenes and terpenoids are key components of essential oils, usually composed of mono- and sesquiterpenes, which may be hydrocarbons or oxygenated compounds. However, these compounds can also be aliphatic or alicyclic derivatives, and some may have aromatic structures.

Terpene compounds are natural products isolated from different natural sources. These are known as secondary metabolites since they are formed by enzymatic reactions of primary metabolites (amino acids, sugars, vitamins), and they consist of five-carbon isoprene units, which are assembled from many isoprene units. Terpenes are simple hydrocarbons, while terpenoids are modified terpenes with different functional groups (ketones, alcohols, esters, carboxylic acids, aldehydes) and oxidized methyl groups moved or removed at various structural positions.

Brazil is considered the country with the greatest biodiversity in the world. According to the Brazilian Ministry of Environment [17], at least 103,870 animal species and 43,020 plant species are known in the country. These species are distributed in six biomes with different characteristics: Amazon Rainforest, Cerrado, Atlantic Forest, Caatinga, Pantanal, and Pampa. Due to its incredible natural wealth, Brazil stands out as one of the leading exporters of essential oils and derivatives.

Brazilian Production of Essential Oils and Derivates

The industrialization of essential oils in Brazil began in the 20th century based on extraction of native essences, basically rosewood. In 1940, due to the great demand by Western industries, deprived of their traditional sources of supply due to the World War II, production of essential oils began in a more organized way with the introduction of other crops to obtain oils from mint, orange, sassafras cinnamon, eucalyptus, lemongrass, patchouli, and others, with a focus on foreign markets [18]. The country enjoys a prominent place in the production of essential oils and is considered one of the four major producers alongside India, China, and Indonesia. The Brazilian position is due to citrus essential oils, obtained as by-products of the juice industry, with limonene being the principal terpene obtained from processing essential oils [19].

Brazilian biodiversity offers excellent potential for producing terpene and terpenoid compounds from essential oils of native flora. Effective coordination between the government, research institutions, and the private sector are needed to obtain quality products and prices capable of competing in the market. ABRAPOE (Brazilian Association of Essential Oils Producers) was founded in 2008 for this purpose. Its objective is to bring together producers and research centers to improve essential oils through research and standardization studies, provide market data and represent the class in government programs [19].

Even though Brazil has the most extraordinary biodiversity, predatory exploitation of natural resources has led to a scarcity of raw materials, including terpenes and terpenoids. A representative case is the collection of the essential oil rich in safrole from cinnamon sassafras. The sources - large trees - have already been sold out in Brazil [20]. Deforestation also offers risks to the production of terpenes and terpenoids from native flora. Recently, it was reported that the destruction of Brazilian biomes increased by 14% in 2020 [21].

Biodiversity, Distribution of Flora, and Distribution of Terpenes and Terpenoids in Biomes

Brazilian flora and fauna are distributed in six terrestrial biomes. Brazil hosting 15-20% of the world's biological diversity, is the most biologically diverse country [17]. The native vegetation covering a region is directly linked to its edaphoclimatic conditions. Brazil has a great diversity of plant species with unique characteristics in each biome.

Table 1 presents Brazilian biome characteristics [17], some native plants and their distribution in the biomes, and some terpenes and terpenoids in essential oils and their pharmacological activities.

Table 1. Brazilian Biomes: Characteristics, Terpenes and Terpenoids from Native Flora

Biome	**Flora**	**Native species**	**Terpene/ Terpenoid**	**Molecular structure**	**Application**
1.Amazon Rainforest	It is the largest forest formation on the planet, conditioned by the humid equatorial climate. Most of Brazil is covered by the forest of Amazon within which there are about 55,000 different species of plants.	*Piper hispidinervum*	α-terpinolene	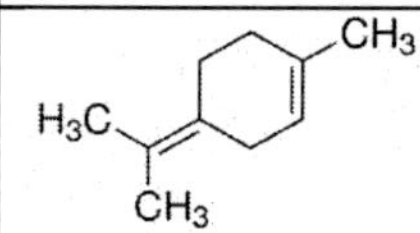	antitumor
2. Cerrado	The Cerrado is made up of grasslands, savannahs, and dry forests. It is the second-largest biome behind the Amazon in South America.	*Lippia gracilis*	thymol	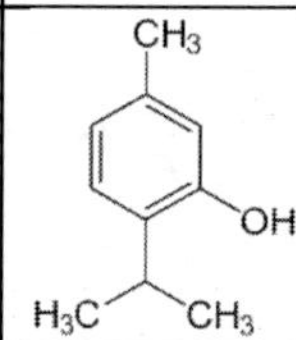	anti-inflammatory
3. Caatinga	The Caatinga has dry soils, and its vegetation is composed of palm trees. This biome is at high risk of desertification.	*Spondias sp*	phytol	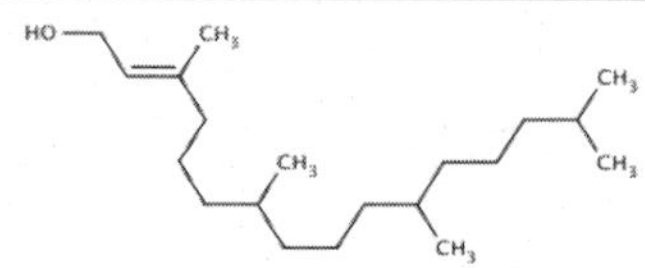	antinociceptive

Biome	Flora	Native species	Terpene/ Terpenoid	Molecular structure	Application
4.Atlantic Forest	The Atlantic Forest has different structures and compositions of flowers due to the climatic characteristics of the region where it occurs, having as a common element the exposure to the humid winds that blow from the ocean.	*Baccharis dracunculifolia*	β-caryophyllene	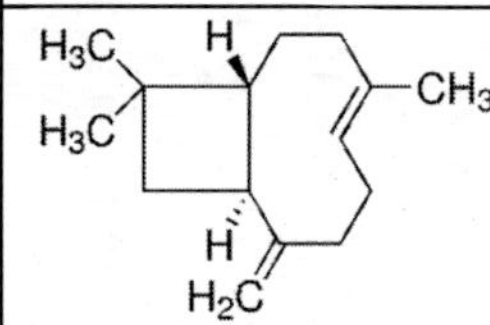	antioxidant
5. Pampa	Herbaceous species characterize this biome. The plains, plateaus, and reliefs have specie of savannah vegetation.	*Bougainvilla glabra*	β-ionone	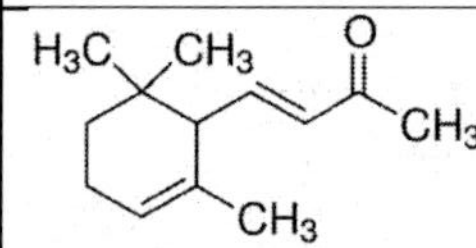	antimicrobial
6. Pantanal	This biome is an alluvial plain influenced by rivers, where fauna and flora of rare beauty and abundance are found. This ecosystem is formed of predominantly sandy terrains, covered by different physiognomies due to variety of microregions and flood regimes.	*Mentha arvensis*	piperitone	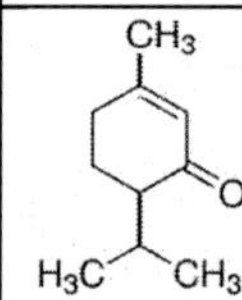	vermifuge

Phytotherapic Uses of Essential Oils and Derivates

The uses of essential oils and their terpenes and terpenoids are highly diverse, depending on source, quality, and extraction procedure. They are widely used in the food, cosmetics, perfumery, cleaning, and pharmaceutical industries. An interesting aspect of these oils is their potential as therapeutic agents. The most common pharmaceutical applications of essential oils are in diluted forms and inhaled in steaming water [22] and as ointments, creams, and compresses [23].

Brazilian flora offers a great diversity of native plants with pharmacological activities associated with essential oils and their derivatives. An example is *Aloysia gratissima*, which has antibacterial, antifebrile, and digestive activities. It is recommended to treat stomach problems, headaches, fevers, bronchitis, pneumonia, flu, and bladder ailments. The essential oil presents sesquiterpenes (α-bisabolol, β-elemene, viridiflorol, and β-caryophyllene), diterpenes (kaurane), and triterpenes (α-amyrin, betulinic acid, and ursolic acid) [24]. The essential oils extracted from Brazilian native plants are still used as an expectorant for treating coughs and bronchitis (*Mikania laevigata*), a regulator of blood pressure (*Psidium cattleianum* Sabine), and a healing agent (*Polygonum punctatum* Elliott), among other uses [24].

Figure 1 presents some Brazilian native plants, their distribution through the biomes, and a terpene or terpenoid compound present in their essential oil. The native plant species represented by numbers 1-6 are linked to Table 1.

Essential Oils Extraction Processes

Traditionally, essential oils are obtained by steam distillation or water distillation from different plant parts [16]. When water and plant material are kept in different recipients, and the water vapor passes through the plant material, contact is brief. The oil can be collected just a few minutes after the process starts [25]. One of the disadvantages of these methods is that essential oils undergo chemical alteration, and the heat-sensitive compounds can easily be destroyed. Therefore, the quality of the essential oil extracts is significantly impaired [26].

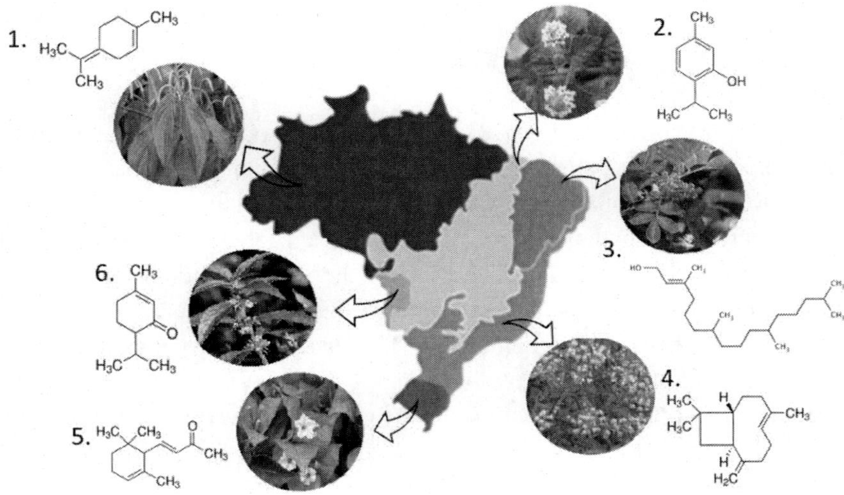

Figure 1. Brazilian native flora: terpenes and terpenoids (authors).

Other processes for obtaining essential oils include maceration, enfleurage, solvent extraction, and use of supercritical fluids. Maceration can be used when the yield from distillation is poor, while enfleurage and solvent extraction are suitable for expensive, delicate, and thermally unstable materials [25].

The extraction of essential oils using supercritical carbon dioxide is more attractive than to conventional processes in terms of product quality [27-29]. Knowledge of the mass-transfer mechanism, the kinetics parameters, and the thermodynamics restrictions of the extraction conducted in a bed of plant material can be used in an economic evaluation of the extraction process [30].

Identification of Terpenes and Terpenoids by Gas Chromatography

The development of chromatographic techniques has contributed considerably to the study of the chemical composition of essential oils. Gas chromatography (GC) is the best method due to its simplicity, rapidity, and efficiency in identifying and quantifying essential oil components and composition variations [25].

Gas Chromatography combined with a flame ionization detector (FID) and mass spectrometry (MS) represents a robust quantitative and qualitative analysis tool. It is also used in quality control, authentication, and detection of adulteration [31]. Additionally, GC is the method of choice for analyzing essential oil compounds [25].

Gas Chromatography coupled with MS makes it possible to carry out in a single operation a qualitative analysis together with an indication of the amount of the components for samples of 1 μL. Kováts indices are standards for comparing substances that are being identified. They have generally been used for GC analyses [32, 33]. In GC, the Kováts indices indicate the compound's retention behavior compared to a mixture of saturated n-alkanes of different numbers of carbon atoms. This parameter provides information about the elution sequence of the compound and varies as a function of stationary phase and temperature, independently of experimental conditions [34]. The retention times are used as references to identify terpene and terpenoid compounds present in essential oils using the Kováts retention indices.

Conclusion

The uses of essential oils are highly diverse, depending on the source, quality, and extraction procedure. An interesting aspect of these oils is their potential as therapeutic agents. These oils have antitumor (*Piper hispidinervum*), anti-inflammatory (*Lippia gracilis*), antinociceptive (*Spondias sp*), antioxidant (*Baccharis dracunculifolia*), antimicrobial (*Bougainvillea glabra*), vermifuge (*Mentha arvensis*) uses, among others.

Brazil has a prominent place in the production of essential oils and their derivatives as terpenes and terpenoids. This is basically due to citrus essential oil with limonene, obtained as a by-product of the juice industry, being the most exported terpene. Brazilian production of essential oils and their derivatives can conquer markets. However, strategic actions such as sustainable extraction, investment in research, and knowledge transfer to producers in cooperatives are needed.

In Brazil, natives have knowledge of the application of essential oils for therapeutical purposes. Thus, it is worth emphasizing the importance of this knowledge to Brazilian culture. With the extinction of these native peoples, their cultures and languages, knowledge about some native plants is also threatened.

European ships brought the measles, smallpox, and flu viruses to Americas. These diseases spread among the natives, invading the coast, and advancing inland. Smallpox decimated the Aztecs and the flu virus, unknown to the natives, caused devastating epidemics, especially among the Guarani. It is estimated that more than 90% of the indigenous population died from the diseases brought by the Europeans [12]. The Brazilian indigenous population witnessed the extermination of several of its people's following encounters with the Europeans. This process, driven by ambition and greed, reduced a population was between two and four million inhabitants in 1500 to around 750,000, spread across Brazil today. In addition to the occupation of its cultural territory, this decline is due to epidemics such as smallpox, measles, whooping cough, chickenpox, typhus, diphtheria, influenza, bubonic plague, among others [35]. However, the wisdom of peoples from the jungles transcends historical time to offer hope to those who claiming to be civilized, especially in times of pandemics.

References

[1] Finchelstein, Federico, Guerisoli, Emmanuel. *Mitos arcaicos del ser Nacional* [*Archaic myths of the National being*]. https://www.clarin.com/opinion/mitos-arcaicos-nacional_0_J7Of4otpq.html. June 28, 2021. Accessed July 13, 2021 (Spanish).

[2] Montoya, Antonio R. The *spiritual achievement made by the religious of the Society of Jesus in the provinces of Paraguay, Paraná, Uruguay, and Tape* (1639). (1997). Porto Alegre: Martins Livreiro Editor. 2nd Ed. 263 p. (Portuguese).

[3] Machado, Alcântara. *Bandeirante's life and death.* (1965) São Paulo: Martins Fontes. 240 p. (Portuguese).

[4] Holanda, Sérgio B. (1994). *The general history of Brazilian civilization: at the colonial time.* v. I, 20 ed. Bertrand Editor. 414 p. ISBN 7803975820658 (Portuguese).

[5] *Brazilian Natives National Foundation (FUNAI).* http://www.funai.gov.br/. Accessed July 2021 (Portuguese).

[6] Braga, Fernão C. (2020). Brazilian traditional medicine: Historical basis, features and potentialities for pharmaceutical development. *Journal of Traditional Chinese Medical Sciences.* https://doi.org/10. 1016/j.jtcms.2020.06.005.

[7] De Mello, Jose F. (1980). Plants in traditional medicine in Brazil. *Journal of Ethnopharmacology*, 49-55.

[8] Pereira, Giovana S., Noelli, Francisca S., Campos, Juliano B., Santos, Marcos P., Zocche, Jairo J. (2016). Guarani historical ecology: the plants

used in the Atlantic Forest biome of the southern coast of Santa Catarina (Part 1). *Cadernos do Lepaarq*, 198-246 (Portuguese). https://periodicos. ufpel.edu.br/ojs2/index.php/lepaarq/ article/view/7608. [*Lepaarq notebooks*]

[9] Guimarães, Veronica M. B. *Belonging to the land: resistance of knowledge and diversity of life by the Kaiowá-Guarani* (2016). PhD diss. University of Brasília, Brasília, 235 p. https://repositorio.unb.br /bitstream/10482/21034/ 1/2016_VeronicaMariaBezerraGuimaraes.pdf Accessed July 2021 (Portuguese).

[10] Bespalez, Eduardo. Archaeological survey and indigenous history in the village of La lima, Miranda/RS, p. 73-91. In: *Indigenous people in Mato Grosso do Sul (2015).* Org. Chamorro, Graciela; Combès, Isabelle. 934 p. https://doi.org/10.30612/9788581471204 (Portuguese).

[11] Chamorro, Graciela. Guarani-speaking indigenous people in the current state of Mato Grosso do Sul (16th-21st centuries) p. 293-322. In: *Indigenous people in Mato Grosso do Sul (2015)* Org. Chamorro, Graciela; Combès, Isabelle. 934 p. https://doi.org/10.30612/ 9788581471204 (Portuguese).

[12] Pavão, Sônia, Lopes, Inair, Vilharva, Kellen N., Pedro, Marildo S., Gislotis, Laura J. (2020) Medicinal plants from the Kaiowa and Guarani peoples as a possible complementary practice in coping with Covid-19 symptoms: traditional knowledge as a weapon against the pandemic. *Brazilian Journal of Agroecology*, 4-17. https://doi.org/10.33240/rba.v15i4.23271 (Portuguese).

[13] Oliveira, Sandra K. S. (2016) *Ethnobotanics in two communities in the indigenous land São Marcos, Roraima, Brazil.* Master's thesis. Federal University of Pará, Belém. 113 p. https://bionorte.propesp. ufpa.br/ ARQUIVOS/teses/VERS%C3%83OFINAL_SANDRA%20KARINY.pdf (Portuguese).

[14] Barbara, Vanessa. (2021). We Knew Bolsonaro Was Guilty. Now We Have 1,288 Pages of Proof. *The New York Times*. https://www. nytimes.com/2021/10/28/opinion/bolsonaro-brazil-report.html. Accessed October 2021.

[15] Ministry of Health Brazil, *Herbal medicine's statistics reports (2016)* (Portuguese) https://www.gov.br/saude/pt-br/assuntos/ noticias/uso-de-fitoterapicos-e-plantas-medicinais-cresce-no-sus/. Accessed September 2021.

[16] Hyldgaard, Morten, Mygind, Tina, Meyer, Rikke L. (2012). Essential oils in food preservation: mode of action, synergies, and interactions with food matrix components. *Frontiers in Microbiology*, 1-24.

[17] Brazilian Ministry of Environment (MMA). *Fourth national report to the convention on biological diversity: Brazil.* Brasília, 2011. 248 p. Acessed August 2021. http://livroaberto.ibict.br/handle/1/965 (Portuguese).
[18] Brazilian Micro and Small Business Support Service (SEBRAE). *How to Build an Essential Oils Factory*, 2009 (Portuguese). https://www.sebrae.com.br/sites/PortalSebrae/ideias/como-montar-uma-fabrica-de-oleos-naturais-e-essencias,c2387a51b9105410Vgn VCM1000003b74010aRCRD#apresentacao-de-negocio. Acessed July 2021.
[19] Bizzo, Humberto R., Hovell, Ana Maria C., Rezende, Cláudia M. (2009). Brazilian essential oils: general view, developments, and perspectives. *Química Nova*, 588-594 (Portuguese).
[20] Bizzo, Humberto R. (2013). Essential oils: resources to be more explored in Brazil. *Lavoura*, 48-53 (Portuguese).
[21] *Report on Greenhouse Gas Emissions Estimation System (SEEG) from the Climate Observatory's (MapBiomas).* (2020). https://alerta. mapbiomas. org/relatorio Accessed September 2021 (Portuguese).
[22] Boehm, Katja, Büssing, Arndt, Ostermann, Thomas. (2012) Aromatherapy as an adjuvant treatment in cancer care - a descriptive systematic review. *African Journal of Traditional, Complementary, and Alternative Medicines*, 503-518.
[23] Karlsen, Jan. (2010). Encapsulation and other programmed release techniques for essential oils and volatile terpenes. Baser, Kemal H. C., Buchbauer, Gehard. (Eds.), *Handbook of Essential Oils. Science, Technology and Applications*, CRC Press, Boca Raton, 855-862.
[24] Garlet, Tanea M. B. *Native medicinal plants of popular use in Rio Grande do Sul.* (2019) Federal University of Santa Maria. 233 p. e-Book. ISBN 978-85-67104-45-4 (Portuguese).
[25] Chamorro, Ester R., Zambón, Silvia N., Morales, Walter G., Sequeira, Alfredo F., Velasco, Gustavo A. (2012) Study of the chemical composition of essential oils by gas chromatography. Salih, Bekir, Çelikbıçak, Ömür. (Eds.), *Gas Chromatography in Plant Science, Wine Technology, Toxicology and Some Specific Applications*, InTech, Rijeka, 307-324.
[26] Illes, Vendel, Daood, Hussein G., Perneczki, S., Szokonya, László, Then, Mária (2000). Extraction of coriander seed oil by CO_2 and propane at super- and subcritical conditions. *The Journal of Supercritical Fluids*, 177-186.
[27] Rodrigues, Vera M., Sousa, Elisa M. B. D., Monteiro, Alcilene R., Chiavone Filho, Osvaldo; Marques, Marcia O. M.; Meireles, Maria Angela A. (2002). Determination of their solubility of extracts from vegetable raw material in pressurized CO_2: a pseudo-ternary mixture formed by cellulosic structure+solute+solvent. *The Journal of Supercritical Fluids*, 21-36.

[28] Rozzi, Nicholas L., Singh, Rakesh K. (2002). Supercritical Fluids and the Food Industry. *Comprehensive Reviews in Food Science and Food Safety*, 33-44.

[29] Ferreira, Sandra R. S., Nikolov, Zivko L., Doraiswamy, Lakshmangudi K., Meireles, Maria Angela A., Petenate, Ademir J. (1999). Supercritical fluid extraction of black pepper (*Piper nigrum* L.) essential oil. *The Journal of Supercritical Fluids*, 235-245.

[30] Pourmortazavi, Seied M., Hajimirsadeghi, Seiedeh S. (2007). Supercritical fluid extraction in plant essential and volatile oil analysis. *Journal of Chromatography A*, 2-24.

[31] Lehotay, Steven L., Hajšlová, Jana. (2002). Application of gas chromatography in food analysis. *TrAC Trends in Analytical Chemistry*, 686-697.

[32] Hoskovec, Michal, Grygarová, Dana, Cvačka, Josef, Streinz, Ludvík, Zima, Jiri, Verevkin, Sergey P., Koutek, Bohumír. (2005). Determining the vapour pressures of plant volatiles from gas chromatographic retention data. *Journal of Chromatography A,* 161-172.

[33] Fonseca, Luciana A. A. P., Sartoratto, Adilson, Cremasco, Marco A. (2021) Experimental determination of thermodynamic properties of terpene and aromatic ketones by gas chromatography *Journal of Molecular Liquids.* doi:10.1016/j.molliq.2020.114531.

[34] Adams, Robert P. (2007) *Identification of essential oil components by gas chromatography/mass spectroscopy*, 4th Ed. Allured Publ. Corp.

[35] Scalco, Nayara, Rodrigues, Eliana. (2013) Changes in the acquisition and consumption of food plants and their relationship with indigenous perceptions of health in a Guarani village, São Paulo, Brazil. *Public Health Nutrition*, 1820-1826.

Biographical Sketches

L. A. A. P. Fonseca

Affiliation: University of Campinas/School of Chemical Engineering

Education: Federal University of Lavras (BS degree)/University of Campinas (Master degree)/University of Campinas (Doctoral degree).

Research and Professional Experience: Thermodynamic properties of bioactive compounds.

Professional Appointments: Professor of Engineering.

Publications from the Last 3 Years:

Papers:

1. Fonseca, L. A. A. P., Cremasco, M. Group contribution methods to predict enthalpy of vaporization of aromatic and terpene ketones at 298.15 K. *Fluid Phase Equilibria*, v. 538, p. 113009, 2021.
2. Fonseca, L. A. A. P., Bertan, A. S., Cremasco, M. A. Group contribution as selection criteria for nutrients fermentation media in the production of tacrolimus. *Chemical.*
3. Fonseca, L. A. A. P., Sartoratto, A. Cremasco, M. A. Experimental determination of thermodynamic properties of terpene and aromatic ketones by gas chromatography. *Journal of Molecular Liquids,* v. 1, p. 114531, 2020.

M. A. Cremasco

Affiliation: University of Campinas/School of Chemical Engineering

Education: State University of Maringa (BS degree)/University of Rio de Janeiro (Master degree)/University of Campinas (Doctoral degree)/University of Purdue (Visiting Scholar)

Research and Professional Experience: Mass transfer processes/Particulate systems

Professional Appointments: Full Professor at University of Campinas/School of Chemical Engineering.

Honors: Unicamp Inventor's Prize (2017, 2019, 2021); Unicamp Education Recognize Prize (2012)

Publications from the Last 3 Years:

Book

1. Cremasco, Marco A. *Mass diffusion.* (In Portuguese). São Paulo: Blucher, 2019. 284 p.

Chapter:

1. Bertan, A. S., Cremasco, M. A. Chemical Engineering and Transdisplinarity against Covid-19. In: Poliana Mendez de Souza e Elizama Aguiar de Oliveira. (Org.). *Engenharia Química: Inovação e tradição em tempos de pandemia* [*Chemical Engineering: Innovation and tradition in times of pandemic*]. 1 ed. Recife: Even 3 Publicações, 2021, v. 1, p. 24-42.

Papers:

1. Cremasco, M. A., Bertan, A. S., Miranda, N. T. Stochastic Diffusion: Walking at Northern Lights. *Jokull*, v. 1, p. 2-14, 2021.
2. Fonseca, L. A. A. P., Cremasco, M. Group contribution methods to predict enthalpy of vaporization of aromatic and terpene ketones at 298.15 K. *Fluid Phase Equilibria*, v. 538, p. 113009, 2021.
3. Fonseca, L. A. A. P., Bertan, A. S., Cremasco, M. A. Group contribution as selection criteria for nutrients fermentation media in the production of tacrolimus. *Chemical Engineering Transactions*, v. 87, p. 577-582, 2021.
4. Bertan, A. S., Silva, S. C. M., Cremasco, M. A. Production of tacrolimus using coconut oil as an alternative to glucose. *Acta Scientiarum. Technology*, v. 43, p. e55721, 2021.
5. Moreira, J. V., Silva, S. C. M., Cremasco, M. A. Evaluation of carbon: nitrogen ratio in semi-defined culture medium to tacrolimus biosynthesis by *Streptomyces tsukubaensis* and the effect on bacterial growth. *Biotechnology Reports*, v. 26, p. e00440, 2020.
6. Bertan, A. S., Cremasco, M. A. Determination of concentration of the pseudomycelles from fermentation broth via spectoscopy and Neubauer chamber. *European Journal of Biological Research*, v. 10, p. 263-270, 2020.

7. Fonseca, L. A. A. P., Sartoratto, A. Cremasco, M. A. Experimental determination of thermodynamic properties of terpene and aromatic ketones by gas chromatography. *Journal of Molecular Liquids,* v. 1, p. 114531, 2020.
8. Bertan, A. S., Cremasco, M. A. The hope as social commitment in the processing of tacrolimus. *The Journal of Engineering and Exact Sciences*, v. 6, p. 0540-0546, 2020.
9. Cremasco, M. A., Taranto, O. P., Castilho, G. J., Tomaz, E. Modelling annual data of PM10 atmospheric in Campinas City from probability density function. *Chemical Engineering Transactions,* v. 74, p. 349-354, 2019.
10. Silva, A. G., Santana, H. S., Bagarolo, R., Rodrigues, A. C., Castilho, G. J., Cremasco, M. A., Taranto, O. P. Application of Raman spectroscopy in microfluidic devices for on-line determination of ethanol concentration in water and vegetable oil. *Chemical Engineering Transactions*, v. 74, p. 739-744, 2019.
11. Silva, S. C. M., Moreira, J. V., Ferrari, W. M. C. S., Franco, T. T., Cremasco, M. A. *Streptomyces tsukubaensis* fermentation using Brazil nut oil to enhance tacrolimus production. *Journal of Applied Biotechnology Reports*, v. 3, p. 109-112, 2019.

Index